The Value of SOCIAL, BEHAVIORAL, and ECONOMIC SCIENCES to National Priorities

A Report for
THE NATIONAL SCIENCE FOUNDATION

Committee on the Value of Social, Behavioral, and Economic Sciences to National Priorities

Division of Behavioral and Social Sciences and Education

A Consensus Study Report of

The National Academies of
SCIENCES · ENGINEERING · MEDICINE

THE NATIONAL ACADEMIES PRESS
Washington, D.C.
www.nap.edu

THE NATIONAL ACADEMIES PRESS   500 Fifth Street, NW   Washington, DC 20001

This activity was supported by Contract No. 10002825 from the National Science Foundation and the Sponsor Award No. SES-1560294. Any opinions, findings, conclusions, or recommendations expressed in this publication do not necessarily reflect the views of any organization or agency that provided support for the project.

International Standard Book Number-13: 978-0-309-45992-1
International Standard Book Number-10: 0-309-45992-3
Digital Object Identifier: https://doi.org/10.17226/24790

Additional copies of this publication are available for sale from the National Academies Press, 500 Fifth Street, NW, Keck 360, Washington, DC 20001; (800) 624-6242 or (202) 334-3313; http://www.nap.edu.

Copyright 2017 by the National Academy of Sciences. All rights reserved.

Printed in the United States of America

Suggested citation: National Academies of Sciences, Engineering, and Medicine. (2017). *The Value of Social, Behavioral, and Economic Sciences to National Priorities: A Report for the National Science Foundation*. Washington, DC: The National Academies Press. doi: https://doi.org/10.17226/24790.

The National Academies of
SCIENCES · ENGINEERING · MEDICINE

The **National Academy of Sciences** was established in 1863 by an Act of Congress, signed by President Lincoln, as a private, nongovernmental institution to advise the nation on issues related to science and technology. Members are elected by their peers for outstanding contributions to research. Dr. Marcia McNutt is president.

The **National Academy of Engineering** was established in 1964 under the charter of the National Academy of Sciences to bring the practices of engineering to advising the nation. Members are elected by their peers for extraordinary contributions to engineering. Dr. C. D. Mote, Jr., is president.

The **National Academy of Medicine** (formerly the Institute of Medicine) was established in 1970 under the charter of the National Academy of Sciences to advise the nation on medical and health issues. Members are elected by their peers for distinguished contributions to medicine and health. Dr. Victor J. Dzau is president.

The three Academies work together as the **National Academies of Sciences, Engineering, and Medicine** to provide independent, objective analysis and advice to the nation and conduct other activities to solve complex problems and inform public policy decisions. The National Academies also encourage education and research, recognize outstanding contributions to knowledge, and increase public understanding in matters of science, engineering, and medicine.

Learn more about the National Academies of Sciences, Engineering, and Medicine at **www.nationalacademies.org**.

The National Academies of
SCIENCES · ENGINEERING · MEDICINE

Consensus Study Reports published by the National Academies of Sciences, Engineering, and Medicine document the evidence-based consensus on the study's statement of task by an authoring committee of experts. Reports typically include findings, conclusions, and recommendations based on information gathered by the committee and the committee's deliberations. Each report has been subjected to a rigorous and independent peer-review process and it represents the position of the National Academies on the statement of task.

Proceedings published by the National Academies of Sciences, Engineering, and Medicine chronicle the presentations and discussions at a workshop, symposium, or other event convened by the National Academies. The statements and opinions contained in proceedings are those of the participants and are not endorsed by other participants, the planning committee, or the National Academies.

For information about other products and activities of the National Academies, please visit www.nationalacademies.org/about/whatwedo.

COMMITTEE ON THE VALUE OF SOCIAL, BEHAVIORAL, AND ECONOMIC SCIENCES TO NATIONAL PRIORITIES

ALAN I. LESHNER (NAM)* (*Chair*), American Association for the Advancement of Science (emeritus), Washington, DC

JOHN S. CARROLL, Sloan School of Management, Massachusetts Institute of Technology

IVY ESTABROOKE, Utah Science Technology and Research Agency, Salt Lake City

RALPH M. GARRUTO (NAS)*, Department of Anthropology, State University of New York, Binghamton

KATHLEEN MULLAN HARRIS (NAS)*, Department of Sociology, University of North Carolina at Chapel Hill

RON HASKINS, Economic Studies and Center on Children and Families, The Brookings Institution, Washington, DC

EDWARD H. KAPLAN (NAE/NAM)*, School of Management, Yale University

RONALD D. LEE (NAS)*, Department of Economics, University of California, Berkeley

ROBERT MOFFITT, Department of Economics, Johns Hopkins University

DUNCAN WATTS, Microsoft Corporation, New York, New York

YANNIS C. YORTSOS (NAE)*, Viterbi School of Engineering, University of Southern California

MELISSA WELCH-ROSS, *Study Director*
NATALIE NIELSEN, *Senior Program Officer*
HOLLY RHODES, *Program Officer*
DIXIE GORDON, *Information Officer*

*NAE, National Academy of Engineering; NAM, National Academy of Medicine; NAS, National Academy of Sciences member.

ACKNOWLEDGMENTS

The committee thanks the project sponsor, the National Science Foundation (NSF), for its support. We particularly thank NSF staff Suzanne Iacono for her efforts in launching this project, and Fay Lomax Cook and Joan Ferrini-Mundy for providing information to the committee both in a public session at the committee's first meeting and as requested during the study process.

At the National Academies of Sciences, Engineering, and Medicine, the committee is extremely grateful for the great leadership, dedication, and competence of the project staff, most notably study director Melissa Welch-Ross, program officer Holly Rhodes, and senior program officer Natalie Nielsen. Their efforts were central to accomplishing our mission. The committee also thanks Sara Frueh and Eugenia Grohman for their helpful assistance with editing and Dixie Gordon and Kelly Arrington for their able handling of administrative matters and meeting logistics.

This Consensus Study Report was reviewed in draft form by individuals chosen for their diverse perspectives and technical expertise. The purpose of this independent review is to provide candid and critical comments that will assist the National Academies of Sciences, Engineering, and Medicine in making each published report as sound as possible and to ensure that it meets the institutional standards for quality, objectivity, evidence, and responsiveness to the study charge. The review comments and draft manuscript remain confidential to protect the integrity of the deliberative process.

The committee thanks the following individuals for their review of this report:

Allison Astorino-Courtois, Chief Analytics Officer and Executive Vice President, NSI, Inc., Lakeway, TX; May R. Berenbaum (NAS)*, Swanlund Professor of Entomology, Department of Entomology, University of Illinois at Urbana–Champaign; Sandra H. Berry, Senior Behavioral Scientist, Senior Director Survey Research Group, and Chair, Human Subjects Protection Committee, RAND Corporation; Brandon L. Garrett, Justice Thurgood Marshall Distinguished Professor of Law, University of Virginia School of Law; Jim Geringer, former Wyoming Governor, Director, Esri, Cheyenne, WY; Jon M. Kleinberg (NAS/NAE)*, Tisch University Professor, Department of Computer Science, Cornell University; Richard G. Kronick (NAM)*, Department of Family Medicine and Public Health, School of Medicine, University of California, San Diego; Bernice A. Pescosolido, Distinguished Professor, Department of Sociology, Indiana University; James M. Poterba (NAS)*, Mitsui Professor of Economics, Department of Economics, Massachusetts Institute of Technology; Richard J. Shavelson, Margaret Jacks Professor of Education and I. James Quillen Dean (emeritus), Stanford University; Jonathan S. Skinner (NAM)*, Department of Economics, Dartmouth College; Harold R. Varian, Chief Economist, Google, Inc., Mountain View, CA; and David H. Wegman, Professor (emeritus), Department of Work Environment, University of Massachusetts, Lowell, and Adjunct Professor, Harvard School of Public Health.

Although the reviewers listed above provided many constructive comments and suggestions, they were not asked to endorse the conclusions or recommendations of this report nor did they see the final draft before its release. The review of this report was overseen by Charles E. Phelps (NAM)*, University Professor and Provost (emeritus), University of Rochester, and Paul R. Gray (NAE)*, Executive Vice Chancellor and Provost (emeritus), University of California, Berkeley. They were responsible for making certain that an independent examination of this report was carried out in accordance with the standards of the National Academies and that all review comments were carefully considered. Responsibility for the final content rests entirely with the authoring committee and the National Academies.

Alan I. Leshner, *Chair*
Committee on the Value of Social,
Behavioral, and Economic Sciences to National Priorities

*NAE, National Academy of Engineering; NAM, National Academy of Medicine; NAS, National Academy of Sciences member.

CONTENTS

Executive Summary 1

The Task 3

    Why Support Research in the Social, Behavioral, and Economic Sciences? 5
    The Core Mission of NSF 8

Do the SBE Sciences Advance NSF's Mission? Health 11

    The Effect of Social Relationships on Health 11
    Disparities in Health and Mortality 12

Do the SBE Sciences Advance NSF's Mission? Prosperity and Welfare 13

    New Ways to Encourage Individuals to Save More for Retirement 13
    Eyewitness Testimony and the U.S. Court System 13
    Bilingualism and Language Development 14
    Willpower and Delay of Gratification 15

Do the SBE Sciences Advance NSF's Mission? National Defense 16

    Terrorism and Counterterrorism 16
    Forecasting Political Instability 16
    Social Network Analysis for the Military and National Intelligence 17

Do the SBE Sciences Advance NSF's Mission? Advancing Progress in Science with Innovative Theories, Methods, and Tools 18

    Groundbreaking Theories of Human Behavior 18
    Understanding How People and Their Circumstances Change Over Time 19
    Simulations, Modeling, and Forecasting 19
    New Methods of Collecting and Analyzing Data 20

Do the SBE Sciences Advance the Missions of Other Federal Agencies? 21

    Auctioning Off Radio Frequencies   21
    Moving from Welfare to Work   21
    Improving National Security, Intelligence, and Counterterrorism   22
    Containing Ebola   23

Do the SBE Sciences Advance the Work of Industry and Business? 24

    Developing Internet Search Engines   24
    Improving Safety in the Airline Industry and in Other Settings   25
    Using the Altitudes of the World Population to Inform Product Development and Marketing   26

Preparing for the Future 27

    A Systematic and Transparent Planning Process   27
    Attention to Trends in Science   28
    Support for Training   33
    Communication of Social, Behavioral, and Economic Research   33

A Final Word 35

Notes 36

EXECUTIVE SUMMARY

Nearly every major challenge the United States faces—from alleviating unemployment to protecting itself from terrorism—requires understanding the causes and consequences of people's behavior. Even societal challenges that at first glance appear to be issues only of medicine or engineering or computer science have social and behavioral components. Having a fundamental understanding of how people and societies behave, why they respond the way they do, what they find important, what they believe or value, and what and how they think about others is critical for the country's well-being in today's shrinking global world. The diverse sciences of the social, behavioral, and economic (SBE) sciences—anthropology, archaeology, demography, economics, geography, linguistics, neuroscience, political science, psychology, sociology, and statistics—all produce fundamental knowledge, methods, and tools that provide a greater understanding of people and how they live.

The National Academies of Sciences, Engineering, and Medicine appointed an expert committee to determine whether the federal government should fund SBE research at the National Science Foundation (NSF), and, specifically, whether SBE research furthers the mission of NSF to advance national priorities in the areas of health, prosperity and welfare, national defense, and progress in science; advance the missions of other federal agencies; and advance business and industry, and to provide examples of such research. The committee was also asked to identify priorities for NSF investment in the SBE sciences from past National Academies reports, if any, and important considerations for NSF for strategic planning.

The committee drew three conclusions based on a review of previous National Academies reports and other research and information from NSF regarding the agency's process for establishing priorities.

CONCLUSION 1 Overall, the social, behavioral, and economic sciences produce a better understanding of the human aspects of the natural world, contributing knowledge, methods, and tools that further the mission of the National Science Foundation to advance health, prosperity and welfare, national defense, and progress in science.

CONCLUSION 2 The understanding, tools, and methods provided by the social, behavioral, and economic sciences—including research supported by the National Science Foundation—provide an essential foundation that helps other agencies achieve their missions.

CONCLUSION 3 The social, behavioral, and economic sciences have provided advances in understanding and tools and methods that have been applicable to business and industry and that enhanced the U.S. economy.

Each of these conclusions is supported by examples of SBE research, many of which have been supported by NSF.

Although it is commendable that NSF consults with advisory groups and with the broader SBE scientific community to identify needs and opportunities in the SBE sciences, in the absence of a strategic plan it is unclear how these inputs are combined and integrated in the agency's SBE priorities. The committee offers four recommendations to better enable SBE research to meet the nation's priorities and challenges.

RECOMMENDATION 1 The National Science Foundation (NSF) should undertake a systematic and fully transparent strategic planning process to provide a clear articulation of the most important scientific questions in the social, behavioral, and economic (SBE) sciences that is consistent with NSF's mission. In addition, NSF's strategic plan should specify the resources and methods required to advance the progress of SBE fields. The plan should reflect broad input from a wide array of stakeholders and put forth priorities for NSF support, while recognizing the need to have a broad and diverse portfolio of innovative projects whose applications may not be immediately apparent but advance the progress of science.

RECOMMENDATION 2 The National Science Foundation (NSF) should continue to support the development of tools, methods, and research teams that can be used to advance the social, behavioral, and economic sciences; facilitate their interactions with other scientific fields; and help NSF and other agencies and organizations more effectively address important national needs.

RECOMMENDATION 3 The National Science Foundation should support training consistent with the ways science is evolving across all scientific fields. Training should prepare the next generation of scientists to be more data-intensive, interdisciplinary, and team oriented.

RECOMMENDATION 4 The National Science Foundation (NSF) should undertake more intensive and systematic efforts to communicate the results and value of the social, behavioral, and economic (SBE) research it supports and how its grants advance NSF's mission. NSF should encourage the broader SBE sciences community to increase its efforts to communicate the results and societal relevance of SBE research.

THE TASK

This report responds to a request from the National Science Foundation (NSF) to the National Academies of Sciences, Engineering, and Medicine (the National Academies) to convene an expert committee and produce a report within 3 months to answer the general question of whether the federal government should fund research in the social, behavioral, and economic (SBE) sciences at NSF. The committee was specifically charged with answering the following questions:

- Do the SBE sciences advance NSF mission areas of national health, prosperity and welfare; securing the national defense; and promoting the progress of science?

- Do the SBE sciences advance the missions of other federal agencies?

- Do the SBE sciences advance the work of industry and business?

- What are priorities for NSF investment in the SBE sciences, and what are important considerations for NSF for future strategic planning?

The committee also was asked to provide examples of how the SBE sciences have helped the nation address societal challenges (see Box 1 for the committee's full statement of task).

The committee interpreted its charge as speaking to the value of SBE research funded by NSF and how it might be strengthened in the future. The committee did not address the question of whether NSF should be funding SBE research, since that is a decision for policy makers, not researchers. Rather, the committee addressed the questions of whether SBE research generally has served NSF's mission areas well and has also served the needs of the nation. Moreover, in the limited time available, the committee did not attempt a comprehensive review of SBE research or even that supported by NSF. Rather, the committee relied on past reports of the National Research Council and the National Academies and the wide-ranging expertise of committee members to identify examples of the contributions of SBE research. Thus, this report contains illustrative examples of SBE studies, some of which have led to great benefits to society and to science, sometimes in surprising ways. The committee mainly identified examples of NSF-funded research but viewed other research as relevant to questions in its charge about whether SBE research has advanced business and industry and national priorities.

> **BOX 1**
>
> ## Statement of Task
>
> The National Academies will appoint an ad hoc committee of approximately eight members to focus on the following:
>
> Should the federal government fund research in the social, behavioral, and economic (SBE) sciences at the National Science Foundation (NSF)? Specific questions to be considered include
>
> - Do the findings, theories, methods, and tools from the SBE sciences in general advance NSF's mission of advancing the national health, prosperity, and welfare; securing the national defense; and promoting the progress of science?
> - Do the SBE sciences advance the missions of other federal agencies?
> - Do the SBE sciences advance the work of industry and business?
> - What priorities for NSF investment, if any, are suggested by prior National Academies reports? What other important elements should be considered as part of a future NSF strategic planning process?
>
> The committee's responses to these questions will provide examples where relevant of how the SBE sciences help the nation address societal challenges, such as the aging of the population, the technological revolution, work and productivity, crime and violence, and the healthy development and well-being of children.
>
> The report is envisioned as a potential resource to solicit comment and input from the broader SBE and constituent communities as part of a more extensive strategic planning process.

After extensive discussion of the research gathered, the committee developed the following **criteria for identifying examples of research** most relevant to answering the charge, although an example did not have to meet all the criteria to be considered:

- The research is of the type that NSF typically funds, more basic than applied.
- The research addresses an issue important to society.
- The research requires minimal detailed technical explanation.
- The research has informed policy or led to discoveries that have advanced national priorities (in health, national defense, welfare and prosperity) or the work of business and industry.
- The research has resulted in broad applications of SBE research to areas not typically associated with the SBE sciences.

- The research findings are counter to common sense, intuition, or generally held beliefs.

- The research has dramatically advanced progress in science or illustrates a trend in science that could lead to significant progress and be applied to advancing national priorities.

These criteria were applied both to research identified in National Academies reports published mainly in the past 20 years and to other research identified by committee members.[1] The committee concluded that the SBE sciences both advance NSF's mission and serve well many of the most important needs of society. The examples provided in this report illustrate how these fields of study can further NSF's mission, the missions of other federal agencies, and the work of industry and business. The committee does not claim that *all* SBE research serves NSF's mission or national needs. As in all fields, the SBE sciences progress through successes and failures. In addition, as noted above, the committee could not do a comprehensive review of all SBE research in the time allotted and thus had to rely on examples. The committee also offers recommendations to improve NSF's strategic planning process in ways that better enable SBE research to meet the nation's priorities and challenges.

WHY SUPPORT RESEARCH IN THE SOCIAL, BEHAVIORAL, AND ECONOMIC SCIENCES?

Every month the Gallup Poll asks a representative sample of Americans "What do you think is the most important problem facing the country today?"[2] The main problems identified include the economy, health care, jobs, race relations, and unemployment. Issues such as these have clear social, behavioral, and economic aspects that need to be better understood, and SBE research can contribute to understanding and addressing them. Moreover, many other problems that at first glance appear to be issues only of medicine or engineering or computer science have social and behavioral components, such as patients' understanding of medical information and community responses to proposed highway development.

Having a fundamental understanding of how people and societies behave, why they respond the way they do, what they find important, what they deeply believe or value, and what and how they think about others is critical in today's shrinking global world. The diverse SBE sciences that are supported at NSF—anthropology, archaeology, demography, economics, geography, linguistics, neuroscience, political science, psychology, sociology, and statistics—produce fundamental knowledge, methods, and tools for a greater understanding of people and how they live. Like all sciences, the SBE sciences bring a rigorous,

> Having a fundamental understanding of how people and societies behave, why they respond the way they do, what they find important, what they deeply believe or value, and what and how they think about others is critical in today's shrinking global world.

methodical approach to pursuing knowledge—collecting data, formulating and testing hypotheses, analyzing evidence—that sheds light on the underlying nature of problems and can help point the way toward remedies. Those remedies depend in part on understanding the social, behavioral, and economic components of problems and how they interact with other aspects.

> Like all sciences, the SBE sciences bring a rigorous, methodical approach to pursuing knowledge—collecting data, formulating and testing hypotheses, analyzing evidence—that sheds light on the underlying nature of problems and can help point the way toward remedies.

Consider, for example, the challenge of immunizing the population against infectious diseases, such as measles and influenza. Medical science has developed many effective vaccines, and when they are administered to the appropriate numbers of people they control the spread of disease. Recent outbreaks of measles, such as those in California and Minneapolis, occurred because not enough parents had their children vaccinated for measles; they did not believe or did not accept the value of vaccination.[3,4] These outbreaks show that individual beliefs and social influences can disrupt vaccination programs and place communities at risk. They also demonstrate that there is a role for the SBE sciences in helping to understand the social and behavioral dynamics of vaccination decisions and using that understanding to develop more effective public health and public information strategies. That is, in addition to the biology of a disease, vaccination efforts require dealing with individuals' and groups' beliefs and decisions about vaccination.

Or consider the task of designing road systems. It may seem to be a relatively straightforward matter, but trying to forecast and understand the decisions that people make about using those road systems play an important role in their design. For example, most drivers will find the shortest possible route to their destination, to minimize their driving time.[5] When a new road is built to alleviate congestion, drivers will take that route if it offers the possibility of less time on the road. But if too many drivers choose the new route, traffic increases and it is no longer faster. This paradox explains why roads that are built to improve traffic flow can quickly become congested[6] and points to the importance of accounting for human preferences and decisions.

Because gaining a complete understanding of many problems and proposing feasible solutions require collaboration between the SBE and other sciences, the SBE sciences are increasingly working with other fields. For instance, meeting many of the challenges recognized in a 2008 report of the National Academy of Engineering, *Grand Challenges for Engineering*, will require collaboration with the SBE sciences.[7] Meeting the challenge to secure cyberspace, for example, will require research on how people interact with computers, the Internet, and information in ways that increase the risk of cybersecurity breaches. Research is also needed to understand the behaviors and social influences on those who commit cybercrimes, such as hackers and saboteurs in organizations.[8] Some interdisciplinary efforts are addressing these issues with combined expertise in business, computer science, economics, law, policy, and social and behavioral sciences.[9]

The numerous contributions of SBE research to society can easily be overlooked—in some cases precisely because the knowledge from that research has become widely accepted. For example, everyone used to think that babies are born as a "blank slate," unable to learn much for the first 6 weeks or so. But research in the early 1970s showed that newborns can learn and remember a variety of associations right away, a fact that people now take for granted.[10]

Another example of now-accepted knowledge comes from the field of polling. Methods and tools for how to ask questions effectively, how to maximize responses, and how to identify representative samples were developed through SBE research.[11] Such understandings and tools are now commonplace in society.

The SBE sciences enable the prediction of many kinds of outcomes with greater certainty, including the success or failure of corporate strategies, economic policies, and legislative agendas.[12,13,14] But while people readily recognize a need for experts from medical research, physics, or biology, when thinking about predicting or explaining human behavior people tend to use "common sense" derived from their own accumulated experiences and anecdotes. Although some people may believe that research knowledge is needed, some who could benefit from SBE research may not be aware that sophisticated tools and insights from the SBE sciences are available to improve understanding and decision making.[15,16,17]

Moreover, common sense can at times be too simple an explanation or just plain wrong. It is commonly believed, for example, that successful people are successful mainly because they are smarter, have worked harder, or are in some other sense more deserving of success than unsuccessful people. However, social science research (including some funded by NSF)[18,19] shows that a large fraction of observed differences in success derive from other factors—such as place of birth, random

> Many leaders of business and industry have long recognized that intuition and common sense are not sufficient, and they use knowledge, tools, and methods from SBE research to understand markets, develop innovations, and inform decisions.

accidents of timing, and the dynamics of competitive markets—that are entirely outside the control of the individuals themselves.[20,21]

Moreover, some findings from SBE research can fail to persuade people precisely because they do not fit with what people already believe to be true. In fact, research in cognitive science has shown that people are more likely to ignore, misremember, forget, or explain away information that does not fit their preconceived notions, such as about how the world works or why people act as they do.[22,23]

Many leaders of business and industry have long recognized that intuition and common sense are not sufficient, and they use knowledge, tools, and methods from SBE research to understand markets, develop innovations, and inform decisions.[24,25,26] Federal, state, and local governments also have begun to recognize the utility of the SBE sciences to both the formulation of policy and the testing of which policies do or do not work in practice.[27]

THE CORE MISSION OF NSF

The federal government has a long history of investment in SBE research. Congress created NSF in 1950 with the unique mission "to promote the progress of science; to advance the national health, prosperity, and welfare; to secure the national defense."[28] Although other, mission-driven federal agencies fund basic research in the SBE sciences—including the National Institutes of Health of the U.S. Department of Health and Human Services and the U.S. Department of Defense—as do foundations, companies, and other organizations, the science that NSF supports often is not directed toward a particular national need or designed to solve a specific problem. Rather NSF-supported basic research is designed to produce foundational understandings on a broad range of topics and develop innovative methods for advancing knowledge. For any type of research, it is not always possible to predict where the research will lead or what effects it will have, but this situation is particularly true for basic research.

Surveys show that people in the United States generally support the federal government funding basic research even though the ultimate uses and effects cannot immediately be known and may take years to unfold. This approach has served the nation well, yielding many benefits both to science and to national priorities.

According to a 2015 survey of U.S. adults, a majority (71%) say government investment in basic science research "pays off in the long run," about the same number of Americans who say engineering and technology "pay off in the long run" (72%). This support for government investments in basic research is unchanged since the questions were last asked in 2009. A majority of adults (61%) consider government funding essential for scientific progress.[29]

The appropriated fiscal 2016 budget for NSF's Social, Behavioral, and Economic Sciences directorate (which funds the largest portion of NSF SBE research) was $272.2 million.[30] The value of providing this basic research funding through NSF, rather than through mission-specific organizations, is that the research is broader in scope and often applies to multiple sciences and to a wide range of problems. For example, the underpinnings of game theory have been applied to research across many global challenges, including kidney transplants, transnational terrorism, and nation-state behavior (see Box 2).

Investments in foundational knowledge from the SBE sciences such as those described in this report have led to many applications and tools that have provided new understanding, ways of addressing societal problems, and enhancements to the quality of life of individuals and for the nation. The stories of such advances sometimes leave the impression that the outcomes were preordained, but such an impression is hindsight. Most often, the findings, methods, and tools that were developed ended up having many different practical uses that were not foreseeable. Even so-called failures in these fields, as in all of science, contribute to knowing which explanations are not accurate and which innovations are unworkable. Just as early technological developments of the microchip took years to develop into computers that were practical to use, research in the fundamental aspects of human behavior can take years to bear practical fruit (see the example in Box 2).

The SBE sciences, like all sciences, pose novel questions and have unique methods. These scientific methods provide systematic ways to gather data through well-designed studies that over time have yielded new understandings of important areas of national interest, as illustrated throughout this report. In this way, SBE research has provided theories, methods, data infrastructure, and tools that are used broadly in the scientific community and beyond.

CONCLUSION 1 The social, behavioral, and economic sciences lead to better understanding of the human aspects of the natural world, contributing knowledge, methods, and tools that further the mission of the National Science Foundation to advance health, prosperity and welfare, national defense, and progress in science.

The next four sections of the report provide discussion and examples of this conclusion in the areas of health; prosperity and welfare; national defense; and advancing progress in science with innovative theories, methods, and tools.

BOX 2 From Game Theory to Kidney Exchange and Many Other Applications

The path from basic research to practical applications can be a long one. Nearly 60 years passed between the earliest research in 1944 on how people behave strategically when playing games[31] and its NSF-supported Nobel Prize-winning application: matching kidney donors to recipients (see figure below). Before this application, economist Alvin Roth used game theory and algorithms to design a fair and efficient way to match medical school internship applicants with hospitals across the country.[32] It turned out that the matching formula had general properties that made it useful for other problems, such as the far more complex situation of finding kidney donors for kidney recipients that were good matches.[33]

The application for kidney matching addressed a common problem: that many people who need a kidney have friends or relatives who are willing to donate, but who turn out to be bad matches for them. Roth developed a computer algorithm that could take into account the thousands of willing donors and thousands of patients needing kidneys, all of whom have different degrees of biological compatibility. His solution matches donor-recipient pairs with other donor-recipient pairs in ways that enable patients to get compatible kidneys—for example, matching donor-recipient pair A with donor-recipient pair B, so that donor A gives a kidney to compatible recipient B, and donor B gives a kidney to compatible recipient A. Newer developments have moved beyond paired kidney exchange to multiperson chains of donors and recipients that reduce patient waiting times for kidney transplants and improve health outcomes.[34] Thousands of kidney transplants have been made possible that otherwise would not have occurred due to this process.[35]

All of these applications evolved from the earliest form of game theory. NSF-funded research building on game theory has led to other Nobel Prize-winning advances in economics[36] and has been applied to many other diverse areas, from the business models of Google, Facebook, and other technology companies to the management of leasing rights for offshore oil fields and forests.[37]

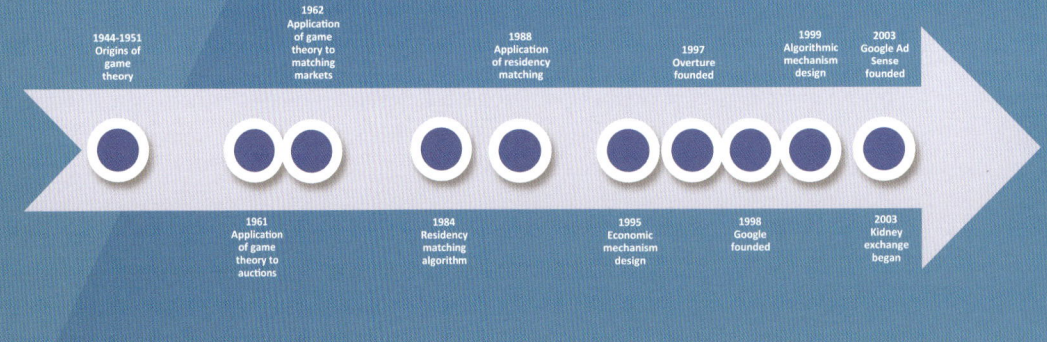

Do the SBE Sciences Advance NSF's Mission? Health

The social, behavioral, and economic (SBE) sciences have played an important role in determining the ways that people's health is affected by a combination of genes and the environment,[38] challenging long-held assumptions about what matters most to health. SBE research has also improved the identification of specific public health needs and ways to promote healthful behavior and prevent illness. Collaborations between SBE researchers and biomedical scientists have identified health problems and early death risks that are disproportionately experienced by large groups of Americans, enabling the development of strategic and timely interventions to improve their health and well-being. Two examples of such important contributions to knowledge include new understandings about how social relationships affect health and how groups of people in the United States differ in their mortality and health.

The Effect of Social Relationships on Health

SBE researchers and neuroscientists working together have found a variety of ways in which behavior and biology affect each other. For example, adversities in life and how people perceive those adversities can determine whether genes are "turned on" ("expressed"), setting up conditions that affect health over the long term. Poverty, violence, and other sources of stress early in life alter children's rapidly developing biological systems[39] and leave them vulnerable to consequences throughout their lives, especially if the children are very young or experience multiple sources of stress—such as abuse and a chaotic home environment—at the same time.[40] These discoveries have informed efforts to prevent child abuse and neglect; there is also good (and growing) evidence that newly developed programs and approaches can help children overcome the effects of these toxic forms of stress.[41]

Research at the intersection of behavior and biology has also revealed that social isolation is a risk factor for early death, comparable in magnitude to well-established risk factors, such as smoking, obesity, and lack of physical activity.[42] In both adolescence and old age, the number and quality of people's social connections have a strong and direct effect on cardiovascular, immune, and metabolic health, resulting in less hypertension, lower rates of inflammation, a lower body mass index (a measure of obesity), and reductions in metabolic syndrome (defined as increased blood pressure, high blood sugar, excess body fat around the waist, and abnormal cholesterol or triglyceride levels).[43]

In adolescence, social isolation increases the risk of inflammation as much as does physical inactivity. In old age, social isolation affects hypertension more than common risk factors, such as diabetes.

Identifying such risks in children before problems occur is critical for intervening early enough to prevent permanent damage to biological systems and avoid the costs they impose on individuals, families, and society.

Disparities in Health and Mortality

Americans differ greatly in health and mortality depending on their social and economic circumstances, and these differences have widened in recent decades. For example, SBE research that compared generations born in 1912 and 1941[44] showed that, between these two generations, life expectancy at age 65 rose by 6 years for people in the top half of the earnings distribution, but by only 1 year for those in the bottom half. This finding is consistent with other studies.[45]

Some groups with lower levels of education, including middle-aged white women without a high school degree, have experienced rising death rates over the last two decades; opioids, suicide, and alcohol-related diseases have played a role in this trend.[46] This and other findings about the health and mortality of the U.S. population provide important data for decisions about public health and government programs, such as Social Security, Medicare, and Medicaid.[47,48]

Do the SBE Sciences Advance NSF's Mission? Prosperity and Welfare

The national need to improve "prosperity and welfare" in the mission statement for the National Science Foundation (NSF) covers a wide range of issues that are of concern to U.S. citizens and that affect the well-being of the nation. Many of these have been the topic of social, behavioral, and economic (SBE) research, such as safe and secure neighborhoods,[49] crime,[50,51] parenting,[52] education,[53,54] the economy, and financial well-being. Four examples are described in this section.

New Ways to Encourage Individuals to Save More for Retirement

Despite the rise of 401(k) and related investments that allow individuals to save through tax-deferred pension plans, employers found that surprisingly few of their eligible employees (only 30%) signed up, opted in, to put any of their salary into those plans, even when their employers matched funds.[55] A dramatic increase—to 90 percent—occurred as a result of a simple change: automatically enrolling workers and then allowing them to opt out rather than requiring them to opt in.[56] Although this change may seem obvious in hindsight, it was informed by research on how people make decisions, process complex information, and think about the future.[57] Since the initial experiments that demonstrated the effectiveness of this approach, many major investment companies that offer retirement plans have adopted it. The research was persuasive enough to lead to 2006 federal legislation requiring firms to make enrollment in such plans the default, that is, to require opting out rather than opting in.

Eyewitness Testimony and the U.S. Court System

Insights from social and behavioral research are shedding light on pitfalls in eyewitness testimony.[58] Using experiments that stage events and ask eyewitnesses to later record their memories, researchers have shown the ways in which eyewitness testimony is fallible. The reliability of eyewitness testimony has also been called into question by the increased use of DNA testing, which has led to reversals of some convictions that had been based primarily on eyewitness testimony.[59]

Basic research on human visual perception and memory—some of which has been supported by NSF—combined with applied research on the factors that affect people who are witnesses to crimes, has illuminated some of the factors that can affect eyewitness testimony. Those factors include low lighting, brief viewing times, large viewing distances, duress, elevated emotions, and the presence of a visually distracting element, such as a weapon; all can affect what witnesses perceive and remember.[60] And perhaps most significantly, people fill gaps in what they see or hear with expectations. Even when incorrect, people often feel certain of their perceptions. Memory also is highly malleable and susceptible to influence. For example, the questions that investigators ask witnesses can affect people's later recall.[61,62,63,64,65,66,67,68] These findings suggest that caution is warranted when using eyewitness procedures in the field and when relying on them in court. There is more research to be done on eyewitness identification, involving scientists, the police, and courts, toward the goal of evidence-based policy and practice.

Bilingualism and Language Development

Many children in the United States—9 percent of the U.S. school-age population—live in homes in which a language other than English is spoken. The parents and teachers of these children often worry that speaking to their child in their home language will interfere with their child's ability to learn English and succeed in school. However, research, including NSF-funded work,[69] is showing that learning two languages either at home or in an early child care setting neither confuses children nor puts them at risk for slower language development.

Indeed, SBE research indicates an underlying human capacity for learning two languages as easily as one.[70] Newborns as young as 0-5 days old can discriminate between the sounds used in different languages. Newborns exposed to English only during their mothers' pregnancies attended more to English sounds than the sounds of the unfamiliar language, while newborns who had been exposed to both languages while in the womb attended to both equally.[71] Over the first year of life, infants become increasingly able to discriminate speech sounds, rhythms, and patterns and to use these early building blocks in their later language development. In addition, adults and children who are competent in two languages may have some cognitive advantages relative to those who only speak one language, such as greater cognitive flexibility, greater ability to regulate behavior, and less cognitive decline at older ages.[72]

Willpower and Delay of Gratification

A person's ability to delay gratification, to exert willpower, at an early age has surprising power to predict important outcomes in school and in life, according to research funded by NSF and others.[73] A simple experiment with hundreds of 4-year-olds more than 40 years ago to study this behavior showed that children differ greatly in their ability to delay gratification.[74] Those differences were shown to have profound effects later in life. After following those children through adolescence and into adulthood, the researchers found that the longer (in seconds) that preschool children could wait for a reward, the higher were their later SAT scores, the better their emotional coping in adolescence, the higher their educational achievement as adults, the lower their rates of substance abuse, and the higher their sense of self-worth.[75]

SBE research also uncovered a set of techniques that can help children delay gratification and control their impulses. These techniques can be taught: when children learned these skills, their long-term outcomes as adolescents and adults were the same as those children who had initially been able to delay gratification.[76]

Do the SBE Sciences Advance NSF's Mission? National Defense

The role of the National Science Foundation (NSF) in securing the national defense largely involves funding some of the basic research that its federal partners—such as the Defense Advanced Research Projects Agency (DARPA), the Army Research Laboratory, the Air Force Office of Scientific Research, the Office of Naval Research and the Naval Research Laboratory, and the U.S. Department of Homeland Security—later use to develop mission-specific tools and applications. Three examples of the contributions of social, behavioral, and economic (SBE) research to national defense are described below. Additional examples are described in the next section of this report, on how SBE research advances the mission of other federal agencies.

Terrorism and Counterterrorism

NSF played an early role in supporting research on terrorism and counterterrorism, beginning in the late 1980s. Recognizing that terrorists' behavior responds to counterterrorism policies in rational ways, this research used game theory (see Box 2) to develop a model to inform counterterror policy.[77] The U.S. Departments of Homeland Security and Defense have sponsored subsequent applications of this research. These applications have focused on government hostage negotiations; an evaluation of the International Criminal Police Organization (INTERPOL) cooperative program for combating international terrorism; the value of counterterror intelligence; and the first evaluation of the use of metal detectors to screen airline passengers.[78]

Forecasting Political Instability

NSF has funded research examining the root causes of political instability and indicators of early observable cues that a change in governance or political state is imminent. That research specifically studied cooperation and conflict, ethnic conflict, state stability, governance, and terrorism.[79] This foundational research was applied through support from DARPA and the U.S. Navy to create the Worldwide Integrated Crisis Early Warning System, which provides policy makers, operational commanders, and intelligence analysts insights and forecasts of changes in political stability.

Social Network Analysis for the Military and National Intelligence

Social network analysis identifies and allows people to understand the relationships among individuals, organizations, and entities.[80] It can illuminate key characteristics of relationships, such as the frequency of communication, affiliation, and other social relationships. Social network analysis can be applied to telephone data, school records, organizational structures, or any other relationship-based data. As an example, some NSF-supported research has combined social network analysis with automated text analysis techniques to provide valuable information about the patterns of behavior of hackers and the vulnerabilities of the nation's cyber networks. These tools analyze hacker chats and other data faster and more efficiently than had previously been possible, with the potential to improve predictions about future threats that are based on hackers' intentions.[81] The U.S. Department of Defense and intelligence agencies have developed additional applications of social network analysis for military and intelligence operations.

Do the SBE Sciences Advance NSF's Mission? Advancing Progress in Science with Innovative Theories, Methods, and Tools

In addition to contributing useful knowledge, the social, behavioral, and economic (SBE) sciences have produced a variety of theories, methods, and tools that are used to predict and explain behavior, identify problems, track them over time, and inform decision making. Social and behavioral research methods also produce rich datasets that are used to test hypotheses about behavior in addition to what can be learned through intuition and experience.

Some of these methods, such as polling or those used to determine the effects of interventions or policies, have become such a part of daily life that it is easy to forget their roots in fundamental SBE research. That research has addressed the complex problems that can complicate the generation of reliable survey data and has developed sophisticated methods for establishing cause-and-effect relationships. In particular, SBE theories and methods have led to a better understanding of conflict and cooperation, and to algorithms that are used for organ-donation matching (see Box 2), predicting international conflict, and modeling crowd behavior. Some of these methods have emerged and advanced with ever-increasing amounts of data and as SBE research addresses new, complex challenges, such as the spread of terrorism.

Groundbreaking Theories of Human Behavior

Although the concept of a theory can be abstract, it has an important role to play in explaining individual and social behavior. Three Nobel Prize-winning bodies of research exemplify some pioneering theories in SBE that have had wide-ranging practical applications.

Game theory explains how individuals reach agreements with one another through conflict or cooperation. It has been enormously influential in the SBE sciences: 11 game theorists have won the Nobel Prize in economics, many of whom received support from the National Science Foundation (NSF). It has been applied to labor markets, industrial organization, arms reduction negotiations, and the provision of public goods (see figure in Box 2).

For a different example, NSF-supported research found an important exception to a prominent social science theory, "the tragedy of the commons." This long-held theory posits that individuals will compete to exploit a public resource, ruining the resource for everyone. One of the implications of this theory was that only a government or governments could limit such individual competition. The new research showed how individuals around the world have cooperated to share resources and develop governance strategies for shared resources that are often better than top-down, government-driven solutions. The principles developed from this work are relevant to current debates about the use of a wide range of collective resources, including the Internet and knowledge in the public domain (e.g., Wikipedia).[82]

Even more recently, the theory of "nudging" describes an approach to policy design that accounts for systematic, irrational tendencies in people's behaviors and decision making,[83,84] building on Nobel Prize-winning work on the psychology of decision making that was funded by NSF.[85] In essence, nudging involves small changes in how choices or options are presented. These near-costless interventions can have remarkable effects. Examples that have yielded demonstrable results include changing the default on organ donation or retirement saving decisions from opt in to opt out (so that no action is required to be in the donation or savings pool) and notifying consumers about their neighbors' energy consumption.[86] The individual and societal benefits of these interventions have been so large that both the U.K. and U.S. governments have established offices dedicated to implementing nudging approaches to a wide range of government programs, with demonstrated policy results.[87]

Understanding How People and Their Circumstances Change Over Time

Studies of education, the labor force, and aging that follow people over long periods of time (longitudinal studies) provide important information about the factors that lead to more or less positive life outcomes. Longitudinal research studies changes in behavior over time and can, in some cases, provide understanding of the long-range outcomes of an intervention. One example is the Health and Retirement Study (HRS) of people aged 50 and older—the premier source of information on the nation's aging population. This large body of data from multiple sciences can help address a wide range of important questions about aging, such as how work, exercise, income, and other factors in middle age, affect circumstances in old age. Many countries around the world have modeled their own surveys after the HRS to understand their own aging populations. This type of research is made possible by access to large datasets—including data from federal statistical agencies and state administrative data systems—and helps to provide a more complete understanding of people and their well-being over time.

Simulations, Modeling, and Forecasting

Models and simulations that apply theories and principles of behavior can be used to develop and test policy ideas and interventions quickly, inexpensively, and safely. One such model of crowd behavior and suicide bombers yielded the surprising finding that remote sensing of suicide bombers and sounding alarms to notify crowds of their presence could actually expose more people to the blast and shrapnel and increase the number of casualties.[88] Models that combine approaches from statistics and demography to more accurately forecast life expectancy and mortality are now widely used around the world by national statistical agencies, public-sector pension agencies, the United Nations, and private-sector providers of life insurance and annuities.[89,90] They are also used by the new longevity swap industry, which helps institutions manage the risks of unknown future costs, such as those for pension plans.[91]

Other examples of NSF-funded models that incorporate SBE research include algorithms that help prevent terrorist attacks[92,93,94] and support sequential decision making to maximize the detection of illicit and hazardous cargo at U.S. ports. A model of pedestrian movement and crowd behavior in dense urban environments[95,96] reveals the rapid exchange of nonverbal information in crowds and shows how the actions of a single individual can shape the dynamics of an entire crowd. Finally, forecasting methods developed by SBE research can now be applied to anonymized and aggregated datasets ("big data") generated by search queries—user browsing logs and social media posts—to predict a wide range of collective human behaviors, such as consumer demand, unemployment claims, and mortgage default rates.[97,98,99]

New Methods of Collecting and Analyzing Data

More and more use is being made of data collected from surveys on the Web and smartphones. However, data collected in this manner do not satisfy a key requirement of standard survey methodology, which is that every member of a given population has to be equally likely to be surveyed. Thus, the data from these surveys are in some sense biased and nonrepresentative. NSF-sponsored research has begun to develop new models that use sophisticated statistical techniques for converting these inherently biased samples into unbiased estimates.[100,101,102] These new methods could dramatically increase the scale, scope, and frequency of obtaining information from survey data by using real-time measures that draw on millions of responses to measure, for example, consumer or business activity, worker productivity, community well-being, or disease caseloads. With more development, these applications have the potential to inform all federal agencies that collect data, including the U.S. Census Bureau, the Bureau of Labor Statistics, the National Center for Health Statistics, the U.S. Office of Management and Budget, and the Centers for Disease Control and Prevention, as well as advance the work of industry and business.

As another example, respondent-driven sampling (also known as network sampling) is a relatively new method that allows researchers to collect important information about "hidden" or hard-to-reach groups, such as those at the greatest risk of infection from HIV/AIDS.[103] This approach relies on members of those groups to recruit each other for the survey. Because data collected in this way are not representative of the total population of infected people, the method also includes statistical procedures for making the data more representative. NSF has supported the further development of respondent-driven sampling,[104,105] and its use has been supported by the U.S. government through the U.S. President's Emergency Plan for AIDS Relief.[106]

Do the SBE Sciences Advance the Missions of Other Federal Agencies?

Social, behavioral, and economic (SBE) research plays a critical role in advancing the missions of federal agencies, such as the U.S. Department of Defense, the U.S. Department of Transportation, the U.S. Department of Health and Human Services, as well as the 17 federal agencies of the nation's intelligence community. These agencies use SBE research and invest in additional, mission-specific research to address their particular needs.

The examples below (as well as those in other sections of this report) describe a range of approaches that the National Science Foundation (NSF) and mission-oriented agencies have taken to address societal, behavioral, and economic components of their mandates. They demonstrate that one fundamental piece of research can affect many different problems, including those being addressed by different agencies, and that one agency's work can be affected by information from many basic research projects.

Auctioning Off Radio Frequencies

The Federal Communications Commission (FCC) sells the radio frequency spectrum to companies that need bandwidth to transmit sound, data, and video to individual and corporate customers. Before the 1990s, the FCC used simple auctions, such as lotteries with random winners, from the list of all bidders. Then, beginning in the early 1990s, the FCC began using research (some of which had been funded by NSF; see Box 2)[107] that had developed mathematical principles to ensure that auction winners would pay a fair price.[108,109] The FCC tested and adopted an algorithm to use with communication companies that allows companies to compete profitably but also ensures that consumers are not overcharged and taxpayers are not subsidizing unreasonable corporate profits. The additional government revenue from the initial auctions has been estimated at $60 billion; because this auction design was adopted by the FCC for later auctions and then spread worldwide, the estimated additional government revenue now totals about $200 billion.[110]

Moving from Welfare to Work

Data from the NSF-funded Panel Study of Income Dynamics (PSID) played an important role in the welfare reform legislation of 1996, which involved multiple agencies, including the U.S. Department of Health and Human Services.[111,112] This ongoing study of a nationally representative sample of families and individuals gathers data on employment, income, wealth, expenditures, health, marriage, childbearing, child development, philanthropy, education, and many other topics. During the welfare reform deliberations, these data were used to determine how and why women moved off of welfare. Counter to the commonly held belief at the time that women left welfare through marriage, the data showed that most women left welfare through work.[113] The PSID also showed that women on welfare worked much more than most people assumed, but that their work was too poorly paid to lift them out of poverty. These findings influenced the inclusion in the welfare reform legislation of work requirements on welfare recipients combined with programs to provide the work-based assistance women needed to care for their families and become self-sufficient, such as child care services, cash income supplements and medical care for mothers and children, transportation assistance, and help for job searches. As a result, there was a significant increase in the number of single mothers who became employed[114] and improved their own economic status and that of their children.[115] However, people have different views about the long-term outcomes of welfare reform based on the same data.[116]

Improving National Security, Intelligence, and Counterterrorism

SBE research originally developed with NSF support—specifically, game theory, social network analysis, development economics, and anthropology—has led to the development of tools and applications that contribute to military capability in current conflicts and the prevention of future conflicts, as well as to efforts to combat terrorism. These capabilities are central to the missions of the U.S. Department of Defense, the intelligence agencies, and the U.S. Department of Homeland Security.

One example is the use of intelligence tools based on social network analysis. As described above (see section on National Defense), NSF funded foundational research on social network analysis[117] that the Office of Naval Research and Air Force Research Laboratory then used to develop a suite of mission-specific tools (e.g., organizational risk analysis and AutoMap). These tools allow analysts to examine key questions, such as, "If this actor is removed from the network, who will likely fill the position in the organizational structure?" The tools have been used in Iraq and Afghanistan to identify key tribal leaders, influential individuals, and the resources available to the networks.[118]

The Worldwide Integrated Crisis Early Warning System (ICEWS), noted above, has leveraged NSF-funded SBE research for tactical, operational, and strategic decision making. ICEWS uses computational models and natural language processing to extract events from newsfeeds and forecast political instability by country. Used by the U.S. Departments of State and Defense and the intelligence community, this system provides data to support policy decisions, understand local conditions, and provide operational insights for commanders. It creates forecasts and data analytic tools to indicate changes in people's behaviors and activities. For example, ICEWS data forecast the 2012 ouster of President Fernando Lugo of Paraguay.

As another example, NSF and the U.S. Department of Defense have both jointly and independently supported research that has addressed the often subtle relationship between conflict, stability, and development. Specifically, analyses of the economic framework that underpins two popular approaches to counterinsurgency and stability operations revealed that ideology is not a primary driver of support for insurgent or terrorist groups.[119] Rather, a population's support for either a government and its allies or terrorist or insurgent groups is more strongly determined by whichever side can best provide public services, such as food, water, safety, and medical care. These findings were taught and used by the U.S. Army and Marine Corps in fighting the insurgency in Afghanistan, as reflected in changes to the Army counterinsurgency manual, and they were used to redefine the way military commanders use funds for rebuilding and reconstruction in that country.

Finally, since 2006 the Marine Corps Center for Advanced Operational Culture and Language has equipped Marines with the cultural skills and information required to function in complex operational environments around the globe. The center's approach is founded on basic social science theory from multiple sciences, among them communication sciences, cultural anthropology, geography, political science, social psychology, and sociology. Marines are taught transferrable concepts and skills that they can use to engage with different cultures. SBE findings, theories, and methods are also used to develop regional and culture-specific content for the Marine Corps' education and training programs, as well as to help the Marine Corps and other U.S. Department of Defense organizations meet challenges associated with stress and resilience and organizational change.

Containing Ebola

Use of anthropological and ethnographic methods played an important role in containing the 2014 Ebola epidemic in Guinea, Liberia, and Sierra Leone. Anthropologists helped to save lives in these nations and to contain the disease, which had the potential to become a global threat. For example, because traditional methods of burial that call for the washing and touching of the dead are believed to have been responsible for 70 percent of new cases of infection in Sierra Leone,[120] anthropologists developed a burial framework that allowed local people to see the body, but not have direct contact with it, and to include burial objects in the body bag prior to burial.[121] The success of anthropologists as mediators in these situations led Médecins Sans Frontières (Doctors Without Borders) to include anthropologists as part of their outbreak response to increase understanding and to facilitate relationships with local populations.[122] Similarly, the World Health Organization is recruiting anthropologists to join its Ebola Virus Disease Outbreak Response Teams to advise on social, cultural, and behavioral factors involved in the spread of Ebola worldwide.[123]

CONCLUSION 2 The understanding, tools, and methods provided by the social, behavioral, and economic sciences—including research supported by the National Science Foundation—provide an essential foundation that helps other agencies achieve their missions.

Do the SBE Sciences Advance the Work of Industry and Business?

Social, behavioral, and economic (SBE) research has had important applications for many fields of business and industry. Social science methods, such as polling, focus groups, forecasting, and statistical modeling, are routinely and widely used to inform consequential business decisions. These decisions relate to all aspects of business, including marketing, customer relations, product development, and strategic planning. Similarly, SBE theories of economic, human, and organizational behavior, including some funded by the National Science Foundation (NSF), have led to the understanding of the Internet as an economic system[124,125,126,127] and influenced business practices across many industries for example, by revolutionizing the pricing of airline tickets, hotel rooms, rental cars, and even cloud computing.[128,129,130,131,132,133,134]

In some cases, basic research has paved the way for new developments, such as Internet search engines. In other cases, theories, principles, and tools from the SBE sciences have been applied to improve operations, reduce accidents, and realize efficiencies in a variety of industries. The first two examples below illustrate some of the complicated ways in which basic research can come to have social and economic value. The third example describes some surprising applications of foundational research on world populations to seemingly unrelated topics.

Developing Internet Search Engines

With a market capitalization of more than $570 billion, Google is the world's second-most valuable company. Google's economic value rests on two pillars: its search engine, which processes over 3.5 billion search queries every day,[135] and its advertising network, which features nearly 30 billion ads per day.[136] Both these capabilities are based on developments in the SBE sciences. Research cited in the Google patent was supported by four different federal science agencies, including an $81,800 grant in 1984 from the sociology program at NSF to study networks of corporate board members.[137,138,139]

The original version of the search engine resulted from a formula developed with NSF funding in the late 1990s by two graduate students. Even in the early days of the Internet, people saw the need for better ways to interact with growing data collections, and early search engines that created indexes of Websites.[140] (Some of these search engines, such as Inktomi and Lycos, also were supported with funding from the Digital Libraries Initiative.) The early researchers recognized that the decision to link pages to each other required conscious effort and the need to reflect human judgment about the significance of the link's destination. This realization led researchers to treat the collection of links as a network, where the "centrality" of a page in the network indicated the page's importance. Using leveraged earlier research by network analysts in mathematics and sociology, researchers created the page rank method, which was the main differentiating feature of the early Google search engine.[141]

The other major breakthrough in Google's development was its use of auctions[142] to set prices for the ads it displays. This advertising model depends heavily on the theory of auctions (discussed above). In 2002, Google used this theory to develop an innovative system that replaced its entire ad sales business with an auction-based platform. Radical at the time, this move formed the foundation of what became Google's profit engine.

Improving Safety in the Airline Industry and in Other Settings

Airline accidents have decreased dramatically over the past 30 years. This reduction is partly due to improved aircraft crew training that is based on fundamental SBE research on team dynamics, leadership, and interpersonal communications. The airline industry used this basic research, in combination with applied research conducted in cockpit simulators and analyses of actual cockpit flight recordings, to develop a training program called crew resource management or cockpit resource management (CRM). CRM is designed so that crew members can communicate effectively and consistently, form an instant team, and adopt well-understood and agreed-upon roles and behaviors.

The National Aeronautics and Space Administration funded some of the early research on CRM, and the National Transportation Safety Board first used the CRM concept in an accident investigation in the late 1970s. United Airlines was the first airline to endorse CRM training, and by the 1990s this approach became a worldwide standard supported by the Federal Aviation Administration and international aviation organizations.[143] Even airlines flying in very different national cultures have incorporated CRM training: South Korea adopted CRM (with the assistance of Boeing trainers) after an airline crash. Many carriers have adapted the CRM training to be consistent with their own cultures. The required behaviors remain the same, but cultural adaptations help those behaviors fit better with patterns of thinking and acting in a given culture, such as when a crew member must speak up to an authority in a culture in which this behavior is generally not accepted.[144,145]

Other industries have copied and tailored CRM techniques. For example, medical schools and hospitals in the United States and around the world now teach and use anesthesia crisis resource management,[146] which draws on the principles of team training. Firefighting crews and emergency responders have also applied CRM principles and training.[147]

More generally, many industries have improved their safety by adopting practices based on SBE research. These practices include work-rest scheduling principles to reduce the fatigue of long-distance truckers in the commercial vehicle industry[148] and the cultivation and assessment of a safety culture in the nuclear power, oil and gas, health care, and other industries.[149] The use of checklists based on CRM principles have spread from airline cockpits to numerous health care settings.[150,151]

Using the Altitudes of the World Population to Inform Product Development and Marketing

Data gathered by SBE scientists as part of an NSF-funded effort to understand the distribution of the world's population by altitude generated unexpected interest from businesses in areas as diverse as food production and packaging, semiconductor manufacturing, and biomedical research and development.[152] In 1998, researchers developed an entirely new mapping technique that divided the earth into grids that were indexed by population size and by altitude.[153] This map revealed that more than one-third of the world's populations lives within 300 feet of sea level and that those populations are distributed in low-density areas, such as agricultural regions.

Many private firms became interested in the findings of this research because of the implications of altitude on several types of products. For instance, Frito-Lay used the data to understand the market for its products at different altitudes because air pressure in packaging needs to be different at different altitudes. Procter & Gamble also had an interest in the altitude distribution because soap and bubbles form differently at various altitudes. Intel was similarly interested because its computer chips cool differently at various altitudes. These applications of the findings of research on the effects of altitudes led to increases in the efficiency or effectiveness of many products.

CONCLUSION 3 The social, behavioral, and economic sciences have provided advances in understanding and tools and methods that have been applicable to business and industry and that enhanced the U.S. economy.

PREPARING FOR THE FUTURE

In response to the question of priorities and elements that the National Science Foundation (NSF) should consider in its strategic plan, the committee recommends attention to four areas that we elaborate below: (1) a systematic and transparent planning process, (2) attention to trends that are occurring in all of science, (3) support for training of researchers and graduate students, and (4) communication of social, behavioral, and economic (SBE) research that NSF supports and how NSF grants advance NSF's mission.

A SYSTEMATIC AND TRANSPARENT PLANNING PROCESS

RECOMMENDATION 1 The National Science Foundation (NSF) should undertake a systematic and fully transparent strategic planning process to provide a clear articulation of the most important scientific questions in the social, behavioral, and economic (SBE) sciences that is consistent with NSF's mission. In addition, NSF's strategic plan should specify the resources and methods required to advance the progress of SBE fields. The plan should reflect broad input from a wide array of stakeholders and put forth priorities for NSF support, while recognizing the need to have a broad and diverse portfolio of innovative projects whose applications may not be immediately apparent but advance the progress of science.

NSF has engaged in extensive and diverse activities to gather input and produce a description of needs and opportunities for research in the SBE sciences in *Rebuilding the Mosaic*.[154] In addition, NSF collects input from internal advisory panels, external advisory panels, and internally managed processes, such as a call for inputs from academic scientists. The committee applauds the fact that NSF consults so widely in forming its priorities. However, the committee found it difficult to determine how these inputs are integrated with one another and how they have influenced the investment portfolio in the absence of a well-articulated strategic plan. Although the process described in *Rebuilding the Mosaic* has identified some important areas for future research, it is not detailed enough to be a strategic plan nor is it described as one.

The committee encourages NSF to go beyond *Rebuilding the Mosaic* to have a clear and publicly available strategic plan that defines SBE priorities, tells how those priorities will be funded, and explains how success in addressing SBE priorities will be evaluated over time. In this process, NSF and the scientific community should articulate the most important scientific questions that will be a priority for investment. It also would be useful for NSF to make available a description of the process by which it incorporates all the input it gathers into the strategic plan. A transparent and inclusive planning process will be needed to explain how broadly or narrowly to define those priorities, weighing national needs and the states of the various SBE sciences.

The strategic planning process should address the unique contributions that the SBE sciences can make to national priorities. Particular attention should be paid to opportunities for SBE research to be integrated with other fields in ways that further understanding of interdisciplinary topics and that address complex societal issues. The planning process should include engagement with other federal agencies that fund SBE research so that all agencies can better specify their unique roles and contributions.

While SBE research has benefited mission-focused agencies, the production of a strategic plan would improve NSF's ability to collaborate with other agencies to ensure that the findings that emerge from the agency's basic research are applied by those agencies and have the maximum impact on improving the lives of Americans. The plan would be highly informative to the SBE scientific communities in planning their own work and research proposals. It would also aid in communicating the findings and implications of the research to the broader scientific community and to the public.

The committee is pleased that *Rebuilding the Mosaic* demonstrates the usefulness of articulating a problem-focused approach to describing the SBE sciences and that this approach illustrates what might be an important direction for NSF for the future. Often, the line between basic and applied research is not clear, and when distinctions can be drawn, each type of research often benefits the other. NSF should consider how progress in basic science may be advanced through research focused on priority scientific questions that have a clear and useful connection to NSF's priority mission areas. (Details of the NSF grant-making process are beyond the committee's charge.)

ATTENTION TO TRENDS IN SCIENCE

RECOMMENDATION 2 The National Science Foundation (NSF) should continue to support the development of tools, methods, and research teams that can be used to advance the social, behavioral, and economic sciences; facilitate their interactions with other science fields; and help NSF and other agencies and organizations more effectively address important national needs.

Some trends in science and society—such as the increasing availability and use of extremely large datasets, the rise of collaborative and interdisciplinary research to address complex problems, and the exponential growth of technology—have implications for future directions of SBE research. Three especially promising directions in *Rebuilding the Mosaic*, along with topics identified in National Research Council and National Academies reports and other sources, are highlighted below for consideration because of their prevalence and importance.

Research Infrastructure, Data Management, Methods, and Measures

The 2017-2021 strategic plan of the Office of Behavioral and Social Science Research of the National Institutes of Health (NIH), in priority #2,[155] states that it is necessary to "enhance and promote the research infrastructure, methods, and measures needed to support a more cumulative and integrated approach to behavioral and social sciences research" and to achieve the following goals:

- A robust and open research infrastructure, including common data elements and consensus measurement metrics.

- Improved precision, accuracy, and efficiency of measures of behavioral and social phenomena and their contexts.

- New methods and analytics to answer increasingly complex research questions relevant to addressing society's most pressing problems.

The committee encourages NSF to adopt the same goals, adjusting them to fit more closely with NSF's mission.

To date, much of the knowledge gained from the social, behavioral, and economic sciences has advanced with data collected and analyzed on a relatively small scale. The committee acknowledges the ongoing importance of small-scale data collection for theory development, but the committee also concurs with previous reports of NSF and NIH that advances in many areas of research increasingly depend on integrating and interpreting data being generated on a much larger scale.

Large-scale datasets are already transforming the questions and methods used in the SBE sciences. These data are being applied to several sectors of society to understand, for example, how people rise out of poverty;[156] improve decision making, performance, and efficiency in health care and many areas of business;[157] understand and prevent cyberattacks;[158] and understand terrorist behaviors.[159,160] However, more progress is needed to advance the use of large datasets in government and across the private sector. For example, survey methodologies that use social media and digital technologies could be developed for real-time measurements that are available instantly and draw on millions of responses to track daily fluctuations in behavior.

Such methods, if well-designed and with careful attention to privacy concerns, could provide a less expensive alternative to household surveys[161] and avoid their limitations in yielding reliable information. *Rebuilding the Mosaic* highlighted a number of data-oriented initiatives for which the SBE sciences community members expressed enthusiasm. Some examples that appear to be worthwhile for enabling and accelerating scientific advances include the following:

- improving access to and making usable existing data for answering scientific questions, such as administrative data collected by federal and state agencies, commercial data, and data that people generate every day as they use social media or electronic devices;

- better integrating the SBE sciences with international data collection efforts and data in other sciences;

- embedding data from social, behavioral, and economic research into geographic information systems and expanding existing geographic information systems to include historical data; and

- transforming analog data (e.g., linguistic and cultural heritage data) to digital form, preferably with geocoding.

Although there is broad enthusiasm for better data infrastructure among the SBE scientific community, the committee notes that the need for such infrastructure is not as clear to many outside that community. The committee therefore suggests that NSF more clearly articulate the important scientific questions—both basic research and applied problems—that depend on the recommended data infrastructure.

As in other areas of research, attention is needed in SBE research to strengthening the evidence base through the replication of research results and to more systematic and transparent documentation and reporting of the circumstances and populations to which the research findings apply.

Fostering Collaborative and Interdisciplinary Research

The committee endorses the conclusion from both *Rebuilding the Mosaic* and the NIH strategic plan noted above that interdisciplinary, team-based research is both closely tied to the development of large-scale data infrastructure and essential for continued progress both in science and when dealing with societal issues related to national priorities. For example, the *Mosaic* report (p. 19) states, "The notion of collaborative research teams is one way that data-intensive SBE research implies a shift away from the independent, single investigator/small team model of scientific research . . . ideas about collaboration, data, technology, and infrastructure are closely intertwined."[162]

NSF should encourage collaborative research at the scale most appropriate for scientific progress, and with attention to what has been learned about the features that create challenges for "team science" and the components of effective approaches.[163] For example, because interdisciplinary work and research teams can be more expensive, take more time, and confront more methodological and organizational hurdles than disciplinary work, incentives are needed to promote such work.

Although funding provides one such incentive, infrastructure support is also important and would increase the impact of grantee funding. Potential examples include coaching or tools to bridge disciplinary boundaries, establish team process, and enhance team functioning; building research networks or consortia; and conferences and other opportunities for researchers to have exposure to such efforts. Scientific progress, however, also depends on simultaneous advances in each of the contributing sciences; thus, increased emphasis on interdisciplinary research should not come at the expense of the core sciences.

Convergence Research

As the examples in this report make clear, the most important problems facing society are complex and interdisciplinary. Their solution depends not only on technological developments but also on understanding their social, behavioral, and economic dimensions. A growing recognition of this complexity, combined with the continuing exponential growth of technology, is leading to a convergence of originally distinct sciences in order to develop new solutions to vexing problems and improvements in people's quality of life. For example, the convergence of engineering and the natural sciences over the past 30 years has contributed to major breakthroughs in multiple areas, far beyond those originally intended. One such area is satellite-based Global Positioning Systems originally developed for military applications that are now part of daily life for many people, ubiquitous in vehicle navigation systems, mobile telephone and tablet computers; as well as in wearables for personalized health monitoring.[164]

In similar ways, SBE research is becoming an increasingly important component of this convergence for solving important problems with global impact.[165,166,167] A 2017 National Academies' report specifically calls for the involvement of social scientists in the conduct of center-based engineering research and the conduct of team science.[168] In addition, the need to understand and address the potentially life-changing effects and unintended consequences of the rapid advance of technology, such as changes in the nature of work as a result of automation, requires the assessment and research of the SBE sciences.[169]

> The most important problems facing society are complex and interdisciplinary. Their solution depends not only on technological developments but also on understanding their social, behavioral, and economic dimensions.

The integration of biology and genetics with SBE research is another example of convergence. SBE research leads to a better understanding of the complex interaction between genes and environments and its effects on health and behavioral outcomes. This understanding combined with technological advances, such as the large-scale collection of DNA and other biomarkers of health risk, has made it possible to predict which groups of people are most at risk for certain physical and mental health problems. It has also led to understanding more about the pathways through which social factors

affect neural and biological processes so that prevention efforts can be more effectively targeted. An example is the Add Health project, a basic science study funded by NIH and 23 other federal agencies, including NSF, to understand how the social environments of young people's lives influence their health and health behavior from adolescence into adulthood. Add Health contains extensive longitudinal information on such health-related behavior as risk taking combined with biological indicators of health, such as obesity, blood pressure, pulse, heart rate, and biomarkers, including DNA.[170]

Another important area in which convergence is advancing understanding is the process of innovation and the growth of innovation ecosystems, a key component of economic development, and the impacts of these ecosystems on individuals and society. That impact is influenced by SBE factors in addition to technical advances. The SBE sciences offer the analysis of ways to accelerate innovation by studying such social dynamics as entrepreneurship, corporate innovation, impact investing, and decision making under uncertainty, as well as public policy. Important SBE topics include the study of the development of regional industrial clusters through entrepreneurship[171] and research on links between patent activities and firm value, identified as a priority research topic by the U.S. Patent and Trademark Office.[172] Studies using both traditional and new methodologies are recognized as a high priority by many organizations, including the National Bureau of Economic Research.[173]

Finally, many additional issues facing American society could benefit from the advances in understanding that the SBE sciences can provide. Practically all emerging

> Practically all emerging technologies, from driverless cars to drones, and the new economic opportunities and other advances they spur, can benefit from the understanding that the SBE sciences can provide about how people desire, experience, and use these technologies, as well as their ultimate effect on people and the larger society.

technologies, from driverless cars to drones, and the new economic opportunities and other advances they spur, can benefit from the understanding that the SBE sciences can provide about how people desire, experience, and use these technologies, as well as their ultimate effect on people and the larger society. Other such challenges include reducing the effects of people's lifestyles on the incidence and progression of diverse diseases; strengthening computer security practices; and managing the economic and labor force impacts that are anticipated with the retirement and aging of the baby boomer generation.

SUPPORT FOR TRAINING

RECOMMENDATION 3 The National Science Foundation should support training consistent with the ways science is evolving across all scientific fields. Training should prepare the next generation of scientists to be more data-intensive, interdisciplinary, and team oriented.

In addition to supporting training in each of the SBE sciences, NSF should support students and researchers in developing the knowledge, skills, and collaborations needed to implement the above recommendations (e.g., for interdisciplinary research, convergence research, team science). Such training has not been central or attended to as yet in a systematic way, as evidenced in a review of the abstracts of the work related to the SBE sciences that the NSF supports, although the need to build such capacity is noted in *Rebuilding the Mosaic*. Consideration needs to be given to the focus of this training, such as whether it should be part of larger efforts to organize and synthesize research or tied to learning specific techniques, such as the skills used in genetics, statistics, or modern neuroscience. NSF has programs to assist undergraduate students, graduate students, and young faculty that could be focused on these recommendations in order to enhance higher education training. The program can also enable midcareer scientists to expand their skills and collaborations with scientists in both other SBE sciences and non-SBE sciences. Training considerations should be part of the strategic planning process in Recommendation 1.

COMMUNICATION OF SOCIAL, BEHAVIORAL, AND ECONOMIC RESEARCH

RECOMMENDATION 4 The National Science Foundation (NSF) should undertake more intensive and systematic efforts to communicate the results and value of the social, behavioral, and economic (SBE) research it supports and how its grants advance NSF's mission. NSF should encourage the broader SBE sciences community to increase its efforts to communicate the results and societal relevance of SBE research.

All NSF research grant proposals are evaluated through peer review that applies two criteria: (1) the scientific merit of the work and (2) its broader societal impacts. The second criterion needs to include communication about the proposed scientific work beyond the scientific community. Such communication should be a high-priority activity that NSF-funded researchers pursue. Describing their work's significance clearly should be expected for the use of all federal funds. To do this well, however, the community of SBE scientists needs to develop or have access to communication expertise to be able to convey the innovations and applications of their research and its value to national priorities. In fact, there is a general need for scientists in all fields to develop sufficient expertise in communication to effectively convey research findings and their impact for the public.

As highlighted throughout this report, the social, behavioral, and economic sciences are important to understanding and intervening to improve virtually every aspect of modern life. Although the value of this research is real and evident to many, it is not as widely understood as it should be. The committee believes that NSF should continue to develop and advance its approaches to communication, including how the research it supports contributes to knowledge on topics of interest and concern to policy makers, federal agencies, journalists, researchers in fields outside the SBE sciences, and the public. Objectives to consider in the communication of the SBE sciences include fostering awareness, understanding, and engagement of diverse stakeholders and enabling the use of SBE research. For example, the use of NSF-funded research to advance health, national defense, and prosperity and welfare relies on the systematic communication of findings to the agencies that can use the results of that research in the design of further research targeted toward each agency's particular mission.

> The committee believes that NSF should continue to develop and advance its approaches to communication, including how the research it supports contributes to knowledge on topics of interest and concern to policy makers, federal agencies, journalists, researchers in fields outside the SBE sciences, and the public.

NSF could play a vital role in providing the resources, training, and tools that scientists need to develop their skills and access the expertise needed to engage in this type of communication. Scientists and scientific institutions also benefit from broad dissemination about new findings, as communication itself promotes scientific progress, spurring others to ask questions that build on that research. Communication approaches should to the greatest extent possible be research informed and assessed for effectiveness in meeting communication goals.

Those approaches need to go beyond the one-way communication of research findings to enhance methods of dialogue with all stakeholders through various forms of public engagement. How best to communicate findings from SBE research and all sciences is itself an important topic for social and behavioral research.

A FINAL WORD

This report highlights some of the many contributions of the social, behavioral, and economic (SBE) sciences to the advancement of knowledge and to meeting national needs. However, it is important to reiterate that the committee could not conduct an exhaustive review and analysis of all SBE research funded by the National Science Foundation (NSF), and, as a result, this report should not be interpreted as implying that all NSF research serves national priorities, furthers the mission of other federal agencies, or advances business and industry. It also should not be interpreted as implying that all SBE research projects are a success, because science, by its very nature, advances through both successes and failures. Moreover, in offering guidance for the future, the committee focused primarily on how the research enterprise can be strengthened, not on offering detailed directions for research. We encourage NSF to engage in a longer, in-depth strategic planning process covering all the relevant content areas to help identify the most productive priorities for research in these fields.

NOTES

[1] Reports of the National Research Council and the National Academies are cited throughout the report as relevant and appear in these endnotes.

[2] See http://www.gallup.com/poll/1675/most-important-problem.aspx [May 2017].

[3] See http://minnesota.cbslocal.com/2017/04/30/vaccine-safety-council-measles-outbreak [May 2017].

[4] See https://www.cdc.gov/mmwr/preview/mmwrhtml/mm6406a5.htm?s_cid=mm6406a5_w [May 2017].

[5] Cohen, H.S. (1995). *Expanding Metropolitan Highways: Implications for Air Quality and Energy.* Transportation Research Board Special Report 245. Washington, DC: National Academy Press.

[6] Steinberg, R., and Zangwill, W.I. (1983). Prevalence of Braess' paradox. *Transportation Science, 17*(3), 301-318.

[7] National Academy of Engineering. (2008). *Grand Challenges for Engineering.* Washington, DC: The National Academies Press.

[8] National Academy of Engineering. (2008). *Grand Challenges for Engineering.* Washington, DC: The National Academies Press.

[9] See http://weis2017.econinfosec.org [May 2017].

[10] Rovee-Collier, C.K. (1997). Dissociations in infant memory: Rethinking the development of implicit and explicit memory. *Psychological Review, 104*(3), 467-498.

[11] Groves, R.M., Fowler F.J., Jr., Couper, M.P., Lepkowski, J.M., Singer, E., and Tourangeau, R. (2011). *Survey Methodology.* Vol. 561. Hoboken, NJ: John Wiley & Sons.

[12] Mellers, B., Stone, E., Murray, T., et al. (2015). Identifying and cultivating superforecasters as a method of improving probabilistic predictions. *Perspectives on Psychological Science, 10*(3), 267-281.

[13] Arrow, K., Forsythe, R., Gorham, M., et al. (2008). The promise of prediction markets. *Science, 320*(5878), 877.

[14] Makridakis, S., Hogarth, R.M., and Gaba, A. (2009). Forecasting and uncertainty in the economic and business world. *International Journal of Forecasting, 25*(4), 794-812.

[15] Lazarsfeld, P.F. (1949). The American soldier—An expository review. *Public Opinion Quarterly, 13*(3), 377-404.

[16] Rosenfeld, S.A. (2011). *Common Sense.* Cambridge, MA: Harvard University Press.

[17] Watts, D.J. (2011). *Everything Is Obvious: *Once You Know the Answer.* New York: Crown Business.

[18] Salganik, M.J., Dodds, P.S., and Watts, D.J. (2006). Experimental study of inequality and unpredictability in an artificial cultural market. *Science, 311,* 854-856.

[19] Salganik, M.J., and Watts, D.J. (2008). Leading the herd astray: An experimental study of self-fulfilling prophecies in an artificial cultural market. *Social Psychology Quarterly, 71,* 338-355.

[20] Frank, R.H. (2016). *Success and Luck: Good Fortune and the Myth of Meritocracy.* Princeton, NJ: Princeton University Press.

[21] Milanovic, B. (2011). *Worlds Apart: Measuring International and Global Inequality.* Princeton, NJ: Princeton University Press.

[22] Nickerson, R.S. (1988). Confirmation bias: A ubiquitous phenomenon in many guises. *Review of General Psychology, 2*(2), 175.

[23] Alba, J.W., and Hasher, L. (1983). Is memory schematic? *Psychological Bulletin, 93*(2), 203.

[24] Wilson, R. (1978). Management and financing of exploration for offshore oil and gas. *Public Policy, 26*(4), 629-657.

[25] Athey, S., and Levin, J. (2001). Information and competition in U.S. Forest Service timber auctions. *Journal of Political Economy, 109*(21), 375-417.

[26] Varian, H.R. (2007). Position auctions. *International Journal of Industrial Organization, 25,* 1163-1178.

[27] See http://38r8om2xjhhl25mw24492dir.wpengine.netdna-cdn.com/wp-content/uploads/2016/10/Behavioral-Insights-for-Cities-2.pdf [May 2017] and http://www.arnoldfoundation.org/initiative/evidence-based-policy-innovation/evidence-based-decision-making [May 2017].

[28] See https://www.nsf.gov/about [May 2017].

[29] See http://www.pewinternet.org/2015/07/01/chapter-3-support-for-government-funding [May 2017].

[30] See https://www.everycrsreport.com/files/20161104R44679_4240907afae9830465342a15697ca5b07a694f23.pdf [May 2017].

[31] Von Neumann, J., and Morgenstern, O. (1944). *Theory of Games and Economic Behavior.* Princeton, NJ: Princeton University Press.

Nash, J. (1951). Non-cooperative games. *The Annals of Mathematics, 54*(2), 286-295.

[32] Roth, A.E. (2003). The origins, history, and design of the resident match. *JAMA, 289*(7), 909-912.

[33] Roth, A.E., Sönmez, T., and Ünver, M.U. (2005). A kidney exchange clearinghouse in New England. *American Economic Review, 95*(2), 376-380.

[34] Roth, A.E. (2015). *Who Gets What—and Why: The New Economics of Matchmaking and Market Design.* New York: Houghton Mifflin Harcourt.

[35] Roth, A.E. (2015). *Who Gets What—and Why: The New Economics of Matchmaking and Market Design.* New York: Houghton Mifflin Harcourt.

[36] See, for example, Nobelprize.org. (2014). *The Prize in Economic Sciences 2012—Advanced Information.* Available: http://www.nobelprize.org/nobel_prizes/economic-sciences/laureates/2012/advanced.html [May 2017].

[37] Personal communication from Hal Varian, professor of economics at the University of California, Berkeley, March 27, 2017.

[38] Hertzman C., and Boyce, T. (2010). How experience gets under the skin to create gradients in developmental health. *Annual Review of Public Health, 31,* 329-347.

[39] Institute of Medicine and National Research Council. (2014). *New Directions in Child Abuse and Neglect Research.* Washington, DC: The National Academies Press.

[40] Hertzman C., and Boyce, T. (2010). How experience gets under the skin to create gradients in developmental health. *Annual Review of Public Health, 31,* 329-347.

[41] Thompson, R.A., and Haskins, R. (2014). Early stress gets under the skin: Promising initiatives to help children facing chronic adversity. *Future of Children, 24*(1), 1-6. Available: http://www.futureofchildren.org/sites/futureofchildren/files/media/helping_parents_helping_children_24_01_policy_brief.pdf [June 2017].

[42] House, J.S., Landis, K.R., and Umberson, D. (1988). Social relationships and health. *Science 241*(4865), 540-545.

[43] Yang, Y.C., Boen, C., Gerken, K., et al. (2016). Social relationships and physiological determinants of longevity across the human life span. *Proceedings of the National Academy of Sciences, 113*(3), 578-583.

[44] Waldron, H. (2007). Trends in mortality differentials and life expectancy for male Social Security-covered workers, by socieoeconomic status. *Social Security Bulletin, 67*(3).

[45] National Academies of Sciences, Engineering, and Medicine. (2015). *The Growing Gap in Life Expectancy by Income: Implications for Federal Programs and Policy Responses.* Washington, DC: The National Academies Press.

[46] Case, A., and Deaton, A. (2015). Rising morbidity and mortality in midlife among white non-Hispanic Americans in the 21st century. *Proceedings of the National Academy of Sciences, 112*(49), 15078-15083.

[47] National Academies of Sciences, Engineering, and Medicine. (2015). *The Growing Gap in Life Expectancy by Income: Implications for Federal Programs and Policy Responses.* Washington, DC: The National Academies Press.

[48] National Academies of Sciences, Engineering, and Medicine. (2015). *The Growing Gap in Life Expectancy by Income: Implications for Federal Programs and Policy Responses.* Washington, DC: The National Academies Press.

[49] National Research Council. (2000). *From Neurons to Neighborhoods: The Science of Early Childhood Development.* Washington, DC: National Academy Press.

[50] National Research Council. (2014). *The Growth of Incarceration in the United States: Exploring Causes and Consequences.* Washington, DC: The National Academies Press.

[51] National Research Council. (2013). *Reforming Juvenile Justice: A Developmental Approach.* Washington, DC: The National Academies Press.

[52] National Academies of Sciences, Engineering, and Medicine. (2016). *Parenting Matters: Supporting Parents of Children Ages 0-8.* Washington, DC: The National Academies Press. doi: 10.17226/21868.

[53] National Research Council. (2000). *How People Learn: Brain, Mind, Experience, and School:*

Expanded Edition. Washington, DC: National Academy Press.

[54] National Research Council. (2007). *Taking Science to School: Learning and Teaching Science in Grades K-8*. Washington, DC: The National Academies Press.

[55] Investment Company Institute (2006) as cited in Thaler, R.H., and Sunstein, C.R. (2008). *Nudge: Improving Decisions About Health, Wealth, and Happiness.* New Haven, CT: Yale University Press.

[56] Thaler, R.H., and Benartzi, S. (2004). Save more tomorrow™: Using behavioral economics to increase employee saving. *Journal of Political Economy, 112*(S1), S164-S187.

[57] Thaler, R.H., and Sunstein, C.R. (2008). *Nudge: Improving Decisions About Health, Wealth, and Happiness.* New Haven, CT: Yale University Press.

[58] National Research Council. (2014). *Identifying the Culprit: Assessing Eyewitness Identification.* Washington, DC: The National Academies Press.

[59] Garrett, B.L. (2012). *Convicting the Innocent: Where Criminal Prosecutions Go Wrong.* Cambridge, MA: Harvard University Press.

[60] Afraz, A., Vaziri-Pashkam, M., and Cavanagh, P. (2010). Spatial heterogeneity in the perception of face and form attributes. *Current Biology, 20*(23), 2112-2116.

[61] Wixted, J.T. (2004). The psychology and neuroscience of forgetting. *Annual Review of Psychology, 55*, 235-269.

[62] Tulving, E., and Thomson, D.M. (1973). Encoding specificity and retrieval processes in episodic memory. *Psychological Review, 80*(5), 352-373.

[63] Dudai, Y. (2006). Reconsolidation: The advantage of being refocused. *Current Opinion in Neurobiology, 16*(2), 174-178.

[64] Loftus, E.F. (2005). Planting misinformation in the human mind: A 30-year investigation of the malleability of memory. *Learning and Memory, 12*(4), 361-366.

[65] Bjork, R.A. (1992). Interference and memory. In L.R. Squire (Ed.), *Encyclopedia of Learning and Memory* (pp. 283-288). New York: Macmillan.

[66] McGeoch, J.A. (1932). Forgetting and the law of disuse. *Psychological Review, 39*(4), 352-370.

[67] Jenkins, J.G., and Dallenbach, K.M. (1924). Obliviscence during sleep and waking. *The American Journal of Psychology, 35*(4), 605-612.

[68] Underwood, B.J., and Postman, L. (1960). Extra-experimental sources of interference in forgetting. *Psychological Review, 67*(2), 73-95.

[69] See, for example, Conboy, B.T., and Kuhl, P.K. (2011). Impact of second-language experience in infancy: Brain measures of first-and second-language speech perception. *Developmental Science, 14*(2), 242-248.

[70] National Academies of Sciences, Engineering, and Medicine. (2017). *Promoting the Educational Success of Children and Youth Learning English: Promising Futures.* Washington, DC: The National Academies Press. doi: 10.17226/24677.

[71] Byers-Heinlein, K., Burns, T.C., and Werker, J.F. (2010). The roots of bilingualism in newborns. *Psychological Science, 21*(3), 343-348.

[72] National Academies of Sciences, Engineering, and Medicine. (2017). *Promoting the Educational Success of Children and Youth Learning English: Promising Futures.* Washington, DC: The National Academies Press. doi: 10.17226/24677.

[73] See https://www.goldengooseaward.org/awardees/marshmallowtest [May 2017].

[74] Mischel, W., Ayduk, O., Berman, M.G., et al. (2011). "Willpower" over the life span: Decomposing self-regulation. *Social Cognitive and Affective Neuroscience, 6*(2), 252-256.

[75] Mischel, W., Ayduk, O., Berman, M.G., et al. (2011). "Willpower" over the life span: Decomposing self-regulation. *Social Cognitive and Affective Neuroscience, 6*(2), 252-256.

[76] See https://www.goldengooseaward.org/awardees/marshmallowtest [May 2017].

[77] Sandler, T., Tschirhart, J.T., and Cauley, J. (1983). A theoretical analysis of transnational terrorism. *American Political Science Review, 77*(4), 36-54.

[78] Enders, W., and Sandler, T. (2012). *The Political Economy of Terrorism, Second Edition.* Cambridge, UK: Cambridge University Press.

[79] Goldstein, J.S. (1992). A conflict-cooperation scale for WEIS events data. *Journal of Conflict Resolution, 36*(2), 369-385.

[80] Wasserman, S., and Faust, K. (1994). S*ocial Network Analysis: Methods and Applications.* New York: Cambridge University Press.

[81] Benjamin, V., and Chen, H. (2012). Securing cyberspace: Identifying key actors in hacker communities. *Intelligence and Security Informatics.* Available: http://www.victorbenjamin.com/papers/

conference/2012/Securing%20Cyberspace%20 Identifying%20Key%20Actors%20in%20 Hacker%20Communities.pdf [May 2017].

[82]See http://evonomics.com/tragedy-of-the-commons-elinor-ostrom [May 2017] and http://www.aei.org/publication/elinor-ostrom-and-the-solution-to-the-tragedy-of-the-commons [May 2017].

[83]Ariely, D. (2008). *Predictably Irrational*. New York: Harper.

[84]Thaler, R.H., and Sunstein, C.R. (2008). *Nudge: Improving Decisions about Health, Wealth, and Happiness*. New Haven, CT: Yale University Press.

[85]Kahneman, D. (2011). *Thinking, Fast and Slow*. New York: Farrar, Straus, and Giroux.

[86]Alcott, H., and Rogers, T. (2014). The short-run and long-run effects of behavioral interventions: Experimental evidence from energy conservation. *American Economic Review, 104*(10), 3003-3037.

[87]See http://www.behaviouralinsights.co.uk [May 2017]; https://sbst.gov [May 2017], and https://sbst.gov/download/2016%20SBST%20 Annual%20Report.pdf [May 2017].

[88]Kaplan, E.H., and Kress, M. (2005). Operational effectiveness of suicide-bomber-detector schemes: A best-case analysis. *Proceedings of the National Academy of Sciences, 102*, 10399-10404.

[89]Lee, R., and Carter, L. (1992). Modeling and forecasting U.S. mortality. *Journal of the American Statistical Association, 87*(419), 659-671.

[90]Gerosi, F., and King, G. (2008) *Demographic Forecasting*. Princeton, NJ: Princeton University Press.

[91]International Monetary Fund. (2012). The financial impact of longevity risk. Chapter 4 in *Global Financial Stability Report. The Quest for Lasting Stability*. Washington, DC: International Monetary Fund. Available: https://www.imf.org/external/pubs/ft/gfsr/2012/01/pdf/c4.pdf [May 2017].

[92]Pita, J., Jain, M., Marecki, J., et al. (2008). Deployed ARMOR protection: The application of a game theoretic model for security at the Los Angeles International Airport. In *Proceedings of the 7th International Joint Conference on Autonomous Agents and Multiagent Systems: Industrial Track* (pp. 125-132). Available: http://teamcore.usc.edu/papers/2008/AAMASind2008Final.pdf [June 2017].

[93]Shieh, E., An, B., Yang, R., et al. (2012). PROTECT: A deployed game theoretic system to protect the ports of the United States. In *Proceedings of the 11th International Conference on Autonomous Agents and Multiagent Systems—Volume 1* (13-20). Available: http://teamcore.usc.edu/papers/2012/protect_aamas_2012_camera_ready_final2_20120109.pdf [June 2017].

[94]An, B., Shieh, E., Tambe, M., et al. (2012). PROTECT—A deployed game theoretic system for strategic security allocation for the United States Coast Guard. *AI Magazine, 33*(4), 96. Available: https://www.aaai.org/ojs/index.php/aimagazine/article/view/2401 [June 2017].

[95]Helbing, D., Farkas, I., and Vicsek, T. (2000). Simulating dynamical features of escape panic. *Nature, 407*(6803), 487-490.

[96]Helbing, D., Buzna, L., Johansson, A., and Werner, T. (2005). Self-organized pedestrian crowd dynamics: Experiments, simulations, and design solutions. *Transportation Science, 39*(1), 1-24.

[97]Choi, H. and Varian, H. (2009). *Predicting Initial Claims for Unemployment Benefits*. Available: https://static.googleusercontent.com/media/research.google.com/en//archive/papers/initialclaimsUS.pdf [May 2017].

[98]Goel, S., Hofman, J.M., Lahaie, S., et al. (2010). Predicting consumer behavior with Web search. *Proceedings of the National Academy of Sciences, 107*(41), 17486-17490.

[99]Choi, H., and Varian, H. (2012). Predicting the present with Google Trends. *Economic Record, 88*(s1), 2-9.

[100]Gelman, A. (2007). Struggles with survey weighting and regression modeling. *Statistical Science, 2007*, 153-164.

[101]Gelman, A., Lax, J., Phillips, J., Gabry, J., and Trangucci, R. (2016). *Using Multilevel Regression and Poststratification to Estimate Dynamic Public Opinion*. Available: http://www.columbia.edu/~jhp2121/workingpapers/MRT.pdf [May 2017].

[102]Wang, W., Rothschild, D., Goel, S., and Gelman, A. (2014). Forecasting elections with non-representative polls. *International Journal of Forecasting, 31*(3), 980-991.

[103]Heckathorn, D.D. (1997). Respondent-driven sampling: A new approach to the study of hidden populations. *Social Problems, 44*(2), 174-199.

[104]Gile, K.J., Johnston, L.G., and Salganik, M.J. (2015). Diagnostics for respondent-driven sampling. *Statistics in Society, 178*(1), 241-269.

[105]Wejnert, C. (2009). An empirical test of respondent-driven sampling: Point estimates, variance, measures of degree, and out-of-equilibrium data. *Sociological Methodology, 39*(1), 73-116.

[106]See https://www.pepfar.gov [May 2017].

[107]Vickrey, W. (1961). Counterspeculation, auctions, and competitive sealed tenders. *The Journal of Finance, 16*(1), 8-37.

[108]Milgrom, P., and Weber, R. (1982). A theory of auctions and competitive bidding. *Econometrica: Journal of the Econometric Society, 50*(5), 1089-1122.

[109]Wilson, R. (1992). Strategic analysis of auctions. In R. Aumann and S. Hart (Eds.), *Handbook of Game Theory with Economic Applications* (Vol. 1, Ch. 8). Amsterdam: North-Holland.

[110]See https://www.goldengooseaward.org/awardees/auction-design [May 2017].

[111]Falk, G. (1994). Welfare: A review of studies on time spent on welfare. *CRS Report for Congress.* Washington, DC: Congressional Research Service.

[112]See http://greenbook.waysandmeans.house.gov/1994-green-book [May 2017].

[113]Harris, K.M. (1993). Work and welfare among single mothers in poverty. *American Journal of Sociology, 99*(2), 317-352.

[114]Acs, G., and Loprest, P. (2001). *Final Synthesis Report of Findings from ASPE's "Leavers" Grants.* Washington, DC: U.S. Department of Health and Human Services.

[115]Haskins, R. (2001). Effects of welfare reform at four years. In P.L. Chase-Lansdale and G. Duncan (Eds.) *For Better and For Worse: Welfare Reform and the Well-Being of Children and Families.* New York: Russell Sage Foundation.

[116]See Burkhauser, R.V., Ed. (2015). Point/counterpoint: Welfare reform: A 20-year retrospective, *Journal of Policy Analysis and Management, 35*(1), 223-244.

[117]Watts, D.J., and Strogatz, S.H. (1998). Collective dynamics of 'small-world' networks. *Nature, 393*(6684), 440-442.

[118]Moon, I-C., and Carley, K.M. (2007). Modeling and simulation of terrorist networks in social and geospatial simensions. *IEEE Intelligent Systems: Special Issue on Social Computing, 22*(5), 40-49.

[119]Berman, E., and Matanock, A.M. (2015). The empiricists' insurgency. *Annual Review of Political Science, 19*, 443-464.

[120]Mumumuza, R. (2014). *Dangerous Practices Spread Ebola in Sierra Leone.* Available: https://www.bostonglobe.com/news/world/2014/12/04/dangerous-practices-spread-ebola-sierra-leone/KWHWTzXQQTYHBXrexLvzXN/story.html [May 2017].

[121]Fassassi, A. (2014). *How Anthropologists Help Medics Fight Ebola in Guinea.* Available: http://www.scidev.net/global/cooperation/feature/anthropologists-medics-ebola-guinea.html [May 2017].

[122]Médecins Sans Frontières. (2014). *Struggling to Contain the Ebola Epidemic in West Africa.* Available: http://www.doctorswithoutborders.org/news-stories/voice-field/struggling-contain-ebola-epidemic-west-africa [May 2017].

[123]World Health Organization, E-Recruitment. (2015). *Ebola Outbreak-Surge Capacity-Anthropologist* (AFRO/14/TA187). Available: https://erecruit.who.int/public/hrd-cl-vac-view.asp?o_c=1000&jobinfo_uid_c=30365&vaclng=en [May 2017].

[124]Varian, H.R. (2016). The economics of Internet search. In J.M. Bauer and M. Latzer (Eds.), *Handbook on the Economics of the Internet.* Northampton, MA: Edward Elgar.

[125]MacKie-Mason, J.K., and Varian, H.R. (1995). Pricing congestible network resources. *IEEE Journal on Selected Areas in Communications, 13*(7), 1141-1149.

[126]MacKie-Mason, J., Shenker, S., and Varian, H.R. (1996). Service architecture and content provision. The network provider as editor. *Telecommunications Policy, 20*(3), 203-217.

[127]Varian, H.R. (2000). Buying, sharing and renting information goods. *The Journal of Industrial Economics, 48*(4), 473-488.

[128]Weatherford, L.R., and Bodily, S.E. (1992). A taxonomy and research overview of perishable-asset revenue management: Yield management, overbooking, and pricing. *Operations Research, 40*(5), 831-844.

[129]Gallego, G., and Van Ryzin, G. (1994). Optimal dynamic pricing of inventories with stochastic

[129] demand over finite horizons. *Management Science, 40*(8), 999-1020.

[130] Talluri, K., and van Ryzin, G. (2004). *The Theory and Practice of Revenue Management.* New York: Springer.

[131] Elmaghraby, W., and Keskinocak, P. (2003). Dynamic pricing in the presence of inventory considerations: Research overview, current practices, and future directions. *Management Science, 49*(10), 1287-1309.

[132] McAfee, R.P., and Velde, V. (2008). Dynamic pricing with constant demand elasticity. *Production and Operations Management, 17*(4), 432-438.

[133] Ebert, P., and Freibichler, W. (2017). Nudge management: Applying behavioral science to increase the knowledge of worker productivity. *Journal of Organization Design, 6*(4), 1-6.

[134] Thaler, R., and Sunstein, C. (2008). *Nudge.* New York: Penguin Books.

[135] See http://www.internetlivestats.com/google-search-statistics [May 2017].

[136] See http://www.business2community.com/online-marketing/how-many-ads-does-google-serve-in-a-day-0322253 [May 2017].

[137] Katz, L. (1953). A new status index derived from sociometric analysis. *Psychometrika, 18*(1), 39-43.

[138] Hubbell, C. (1965). An input–output approach to clique identification. *Sociometry, 28,* 377-399.

[139] Mizruchi, M.S., Mariolis, P., Schwartz, M., et al. (1986). Techniques for disaggregating centrality scores in social networks. *Sociological Methodology, 16,* 26-48.

[140] Some of these search engines, such as Inktomi and Lycos, also were supported with funding from the Digital Libraries Initiative.

[141] This paragraph drawn from the two summaries of NSF awards: https://www.nsf.gov/news/special_reports/cyber/digitallibraries.jsp [May 2017] and https://www.nsf.gov/discoveries/disc_summ.jsp?cntn_id=100660 [May 2017].

[142] See https://www.wired.com/2009/05/nep-googlenomics [May 2017].

[143] Helmreich, R.L., and Foushee, H.C. (2010). Why CRM? Empirical and theoretical bases of human factors training. In B.G. Kanki, R.L. Helmreich, and J. Anca (Eds.), *Crew Resource Management* (Ch. 1, pp. 3-58). Cambridge, MA: Academic Press.

[144] Helmreich, R.L., and Foushee, H.C. (2010). Why CRM? Empirical and theoretical bases of human factors training. In B.G. Kanki, R.L. Helmreich, and J. Anca (Eds.), *Crew Resource Management* (Ch. 1, pp. 3-58). Cambridge, MA: Academic Press.

[145] Helmreich, R.L., Merritt, A.C., and Wilhelm, J.A. (1999) The evolution of crew resource management training in commercial aviation. *The International Journal of Aviation Psychology, 9*(1), 19-32. doi: 10.1207/s15327108ijap0901_2.

[146] Gaba, D.M., Howard, S.K., Fish, K.J., et al. (2001). Simulation-based training in anesthesia crisis resource management (ACRM): A decade of experience. *Simulation & Gaming, 32*(2), 175-193.

[147] Hagemann, V., Kluge, A., and Greve, J. (2012). Measuring the effects of team resource management training for the fire service. *Proceedings of the Human Factors and Ergonomics Society Annual Meeting, USA, 56,* 2442-2446. doi:10.1177/1071181312561497.

[148] National Academies of Sciences, Engineering, and Medicine. (2016). *Commercial Motor Vehicle Driver Fatigue: Long-Term Health, and Highway Safety.* Washington, DC: The National Academies Press.

[149] National Academies of Sciences, Engineering, and Medicine. (2016). *Strengthening the Safety Culture of the Offshore Oil and Gas Industry.* Washington, DC: The National Academies Press.

[150] Pronovost, P., and Vohr, E. (2010). *Safe Patients, Smart Hospitals: How One Doctor's Checklist Can Help Us Change Health Care from the Inside Out.* New York: Plume.

[151] Gawande, A. (2007). The checklist. *The New Yorker, 83*(39), 86-95.

[152] See https://www.goldengooseaward.org/awardees/of-maps-and-men [May 2017].

[153] Cohen, J.E., and Small, C. (1998). Hypsographic demography: The distribution of human population by altitude. *Proceedings of the National Academy of Sciences, 95*(24), 14009-14014.

[154] U.S. National Science Foundation. (2011). *Rebuilding the Mosaic: Fostering Research in the Social, Behavioral, and Economic Sciences at the National Science Foundation in the Next Decade.* Directorate for Social, Behavioral, and Economic Sciences. Washington, DC: National Science Foundation.

[155] National Institutes of Health. (2017). *The Office of Behavioral and Social Sciences Research*

[155]*Strategic Plan 2017-2021*. Available: https://obssr.od.nih.gov/wp-content/uploads/2016/12/OBSSR-SP-2017-2021.pdf [May 2017].

[156]Chetty, R., Hendren, N., Kline, P., et al. (2014). Is the United States still a land of opportunity? Recent trends in intergenerational mobility. *The American Economic Review, 104*(5), 141-147.

[157]See https://www.forbes.com/sites/bernardmarr/2015/04/21/how-big-data-is-changing-healthcare/2 [May 2017].

[158]Benjamin, V., and Chen, H. (2012). Securing cyberspace: Identifying key actors in hacker communities. *Intelligence and Security Informatics,* 24-29. Available http://www.victorbenjamin.com/papers/conference/2012/Securing%20Cyberspace%20Identifying%20Key%20Actors%20in%20Hacker%20Communities.pdf [May 2017].

[159]See https://www.bloomberg.com/news/articles/2016-10-13/predicting-terrorism-from-big-data-challenges-u-s-intelligence [May 2017].

[160]Strang, K.D., and Sun, Z. (2016). Analyzing relationships in terrorism big data using Hadoop and statistics. *Journal of Computer Information Systems, 56*(6), 55-65.

[161]Choi, H. and H. Varian. (2012). Predicting the present with Google Trends. *Economic Record, 88*(s1), 2-9.

[162]U.S. National Science Foundation. (2011). *Rebuilding the Mosaic: Fostering Research in the Social, Behavioral, and Economic Sciences at the National Science Foundation in the Next Decade*. Directorate for Social, Behavioral, and Economic Sciences. Washington, DC: National Science Foundation.

[163]National Research Council. (2015). *Enhancing the Effectiveness of Team Science*. Washington, DC: The National Academies Press.

[164]National Research Council. (2014). *Convergence: Facilitating Transdisciplinary Integration of Life Sciences, Physical Sciences, Engineering, and Beyond*. Washington, DC: The National Academies Press.

[165]See https://www.nsf.gov/awardsearch/showAward?AWD_ID=1216048 [May 2017].

[166]Brummitt, C.D., Barnett, G., and D'Souza, R.M. (2015). Couples catastrophes: Sudden shifts cascade and hop among interdependent systems. *Journal of the Royal Society Interface, 12,* 1-22.

[167]Vijayaraghavan, V.S., Noël, P-A., Maoz, Z., et al. (2015). Quantifying dynamical spillover in co-evolving multiples networks. *Scientific Reports, 5*(15142), 1-10.

[168]National Academies of Sciences, Engineering, and Medicine. (2017). *A New Vision for Center-Based Engineering Research*. Washington, DC: The National Academies Press.

[169]National Academies of Sciences, Engineering, and Medicine. (2017). *Information Technology and the U.S. Workforce: Where Are We and Where Do We Go from Here?* Washington, DC: The National Academies Press. doi: 10.17226/24649.

[170]See http://www.cpc.unc.edu/projects/addhealth [May 2017].

[171]Feldman, M., Francis, J. and Bercovitz, J. (2005). Creating a cluster while building a firm: Entrepreneurs and the formation of industrial clusters. *Regional Studies, 39,* 129-141. Available: http://dx.doi.org/10.1080/0034340052000320888 [May 2017].

[172]Graham, S.J.H., and Hancock, G. (2014). The USPTO economics research agenda. *Journal of Technology Transfer, 39,* 335-344.

[173]See http://www.nber.org/callforpapers/CallforProposalsProductivityInnovationEntrepreneurship.html [May2017].

> **Of Interest....**
>
> **From Research to Reward: A National Academy of Sciences Series About Scientific Discovery and Human Benefit** includes six true stories that demonstrate how advances in the social and behavioral sciences often lead to surprising and remarkable benefits for society. These narratives and accompanying videos can be found at www.nasonline.org/r2r.